D1578004

PLUMBING LICENSING

Exam Guide

THIRD EDITION

By Chris Prince

Published by:

DELMAR
CENGAGE Learning™

Australia • Brazil • Japan • Korea • Mexico • Singapore • Spain • United Kingdom • United States

dewalt.com/guides

DELMAR
CENGAGE Learning™

**DeWALT Plumbing Licensing Exam Guide,
Third Edition**
Chris Prince

Vice President, Technology and Trade
Professional Business Unit: Gregory L. Clayton

Director of Building Trades: Taryn Zlatin
McKenzie

Product Development Manager: Robert Person

Development: Nobina Chakraborti

Director of Marketing: Beth A. Lutz

Marketing Manager: Marissa Maiella

Production Director: Carolyn Miller

Production Manager: Andrew Crouth

Production Service/Compositor:
PreMediaGlobal

Art Director: Benjamin Gleeksman

For product information and technology assistance, contact us at
Cengage Learning Customer & Sales Support, 1-800-354-9706

For permission to use material from this text or product,
submit all requests online at **www.cengage.com/permissions.**
Further permissions questions can be e-mailed to
permissionrequest@cengage.com

Library of Congress Control Number: 2010929691

ISBN-13: 978-1-111-13552-2

ISBN-10: 1-111-13552-5

Delmar
5 Maxwell Drive
Clifton Park, NY 12065-2919
USA

Cengage Learning is a leading provider of customized learning solutions with
office locations around the globe, including Singapore, the United Kingdom,
Australia, Mexico, Brazil, and Japan. Locate your local office at:
international.cengage.com/region

Cengage Learning products are represented in Canada by Nelson Education, Ltd.

Visit us at **www.InformationDestination.com**

For more learning solutions, please visit our corporate website at
www.cengage.com

Notice to the Reader
Cengage Learning and DeWALT do not warrant or guarantee any of the products described herein or perform
any independent analysis in connection with any of the product information contained herein. Cengage
Learning and DeWALT do not assume, and expressly disclaim, any obligation to obtain and include information
other than that provided to it by the manufacturer. The reader is expressly warned to consider and adopt
all safety precautions that might be indicated by the activities described herein and to avoid all potential
hazards. By following the instructions contained herein, the reader willingly assumes all risks in connection
with such instructions. Cengage Learning and DeWALT make no representations or warranties of any kind,
including but not limited to, the warranties of fitness for particular purpose or merchantability, nor are any
such representations implied with respect to the material set forth herein, and Cengage Learning and DeWALT
take no responsibility with respect to such material. Cengage Learning and DeWALT shall not be liable for any
special, consequential, or exemplary damages resulting, in whole or part, from the readers' use of, or reliance
upon, this material.

Printed in the United States of America
2 3 4 5 6 7 14 13 12 11

CONTENTS

DeWALT PLUMBING LICENSING EXAM GUIDE—THIRD EDITION

by
Chris Prince

INTRODUCTION

Congratulations! If you are reading this book, it is likely you are pursuing a new opportunity. Whether you are preparing to take an exam in order to start your own plumbing business, to enhance the qualifications of an existing company, or to qualify the company for which you now work, you are taking deliberate measures to improve an existing situation. So sit back, relax, and rest assured, the time you spend with this book will significantly improve your chances of passing the exam you are about to take.

TIME FOR TEST MODE

Preparing for and taking an exam requires a specific mindset. It requires planning for the upcoming event and making a conscious effort to control your anxiety. For most, the mere thought of taking an exam will cause at least a feeling of butterflies churning in the stomach. Some may even react more severely. Don't fear, controlled test anxiety is good. It creates a sense of urgency to succeed. Identify your level of anxiety and control it.

How do you rate your level of anxiety? For some, perhaps the last test-taking ordeal was a high school final exam many years ago. The memory of this experience, coupled with the pressure of living up to the expectations of family, boss, or coworkers, can be overwhelming. When children are aware that a parent will soon be taking an exam, they are likely to reciprocate the pressure applied to do well in school. Bosses often assume the exam will be a breeze for such a competent employee. Many times, a spouse's support is mistaken for unrealistic expectation. Peer pressure is not necessary. Do what you can to minimize it.

The pressures felt from the need to please and live up to the expectations of everyone around you can lead to undue stress and are not necessary. Simply keep the fact that you are taking the exam a secret! Maybe everyone notices your study efforts but there is no need to announce the test date. Make your plans by scheduling the exam, and if you must explain, simply say you have an appointment. If no one knows your plans, the only pressure you will feel to succeed is your own. This pressure is healthy and probably necessary. Rather than dealing with the pressure of others, imagine the feeling when you announce your accomplishment after the fact. Work toward this goal and visualize success. Think positive.

Another way to eliminate stress may be to think of your first test date as a practice run. The worst-case scenario is that passing the exam will take two attempts. The simple fact of the matter: It is not the end

of the world if you do not pass on the first attempt. Most states and municipalities allow you to take the exam as many times as necessary without penalty. Make sure you are not limited by the number of times you can take an exam before using this suggestion. (We will explain later how to check this.)

Next, avoid becoming irritated that the exam is required. Forget the fact that you know your trade and feel you should not be required to pass an exam to prove yourself. Look at the experience as an opportunity. The knowledge you gain from preparing for the exam will far outweigh the effort. Without a doubt, you will gain information that will improve your efficiency and trade expertise and could even save you money.

Remember the old carpenter's advice to measure twice and cut once? Similarly, when it comes to test-taking, it is critical to be precise, careful, and methodical as you prepare for exam day. Apply this philosophy as you take the practice exams in this book—read the question twice and choose the answer carefully. The exams are typically administered in a multiple-choice format and it is not uncommon for at least two of the choices to appear correct, depending on how the question is read. Remember, there is only one correct answer. One word in the question can change the entire meaning of the question. Be careful.

After you have switched to test mode, it is time to begin preparing for the exam.

PREPARING FOR THE EXAM

The first step of preparation is to identify the exam required to accomplish your goal. If you are taking a state-required exam to become a licensed contractor, begin by visiting www.becomealicensedcontractor.com. Select your state, choose plumbing under the listing of test types, and the test requirements will be found. The site will likely lead you to the candidate bulletin provided by the approved testing company.

The candidate bulletin is the test-taker's rulebook. It provides specific information including the addresses of testing centers, identification requirements, security procedures, and anything else necessary for taking the exam. In preparing for the exam, it is important to identify the answers to the following questions using the candidate information bulletin:

1. **What are the approved references?** This is, by far, the most important piece of information to know prior to beginning your exam preparation mission. Most plumbing exams are based on

Notes

both the International Plumbing Code and the International Gas Code. In addition, many exams include content that will be found in an OSHA Standards Part1926 book, a plumbing theory book, and a project management book. After identifying the approved reference materials, you can eliminate the study exams in this book that are not included in your exam. Do not spend valuable time studying information that is not included on your exam.

2. **Are the approved references allowed into the examination center?** In most cases the books are allowed in the examination room to be available to you throughout the exam. While it would seem to make the exam extremely easy, if you are not familiar with how to use the books, it might as well be a closed-book exam.

3. **Is tabbing and highlighting of the books allowed?** Typically, if the approved references are allowed to be taken into the exam room, they are allowed to be highlighted and tabbed prior to test day. Many rules allow you to underline but prohibit you from making notes in the reference materials. The candidate bulletin will specifically call for permanent tabs. Permanent tabs are those which cannot be easily removed. Post-It® tabs are generally not allowed. If any of the rules are broken, your reference materials can be banned from the test site.

4. **What is the time frame for taking the exam?** This is very important because you will want to simulate the time allowed as you work through the study exams throughout this book. Simply take the allotted time in minutes and divide by the number of questions on the exam to determine the time allotment per question. You can now easily multiply the number of questions in a particular section of this book by the average time allowed per question to set the time frame for working through the practice exams. In the beginning, you will want to allot additional time until you become familiar with the reference materials.

5. **What is the content outline of the exam?** It is important to pay attention to the breakdown of the exam for the purpose of allotting your study time. If only 5 out of 80 questions pertain to OSHA, you should spend less time studying this subject and concentrate more on the areas representing the largest portion of the exam.

USING THIS EXAM GUIDE

Before starting the exam preparation process using this guide, make sure you have highlighters and tabs readily available. Begin by identifying the table of contents, the index, and the glossary of each

book. Place a tab on each of these sections for easy access. The table of contents will divide the book into chapters or subjects and will be used frequently. The index is an alphabetical listing of key words found throughout the book and should be your starting point for finding an answer to a specific question. If the answer is not found using the index, identify the chapter according to the subject of the question using the table of contents. The glossary is an alphabetized listing of terms and can serve as a useful source for answering questions quickly. A glossary may or may not be found in the approved references.

Each section of this exam guide is based on a particular plumbing subject, general project management information, OSHA regulations, or a chapter of the *International Plumbing Code* or *International Gas Code* and is so identified. As you answer each of the questions from the study exam, highlight the answers. This will help you become familiar with the book and will strengthen your ability to quickly reference important code items. Remember, the purpose of each practice exam is not to test your trade knowledge, it is to provide an exercise of how to navigate and use the approved reference(s). If you feel you know the answer to the question, you should verify the answer using the applicable book. Pay very close attention to the tables in the references and the subject matter of each chapter.

When it comes to tabbing your reference materials, be careful not to overdo it. Placing too many tabs throughout your reference materials will serve more as a hindrance than a useful tool. Remember, when using the tabs, the more tabs you use, the longer it will take to read through each one. Using the index will likely save you time in your search for specific information. It is highly recommended that you tab the chapters, table of contents, index, and useful tables.

Going into the exam, make sure you are familiar with the subject matter of each book. It is imperative that you understand the layout and format of each of the references. This is accomplished through the process of answering the questions in this guide.

Prior to exam day, remove the index in any of the approved reference books bound with a three-ring binder, staple it together, and place it in the front pocket of the book. This will allow easy access to the index by being able to place it on the testing table beside the reference rather than requiring you to flip back and forth between the index and the body of the book.

On average, exams allow about 3 minutes per question. When you begin using the study exams in this book, allow yourself between

Notes

Notes

8 and 10 minutes per question. As you become more familiar in your use of the reference materials, decrease this time allotment to about 5 minutes. By the time you complete the final exams, allow yourself only the time allotted on your exam.

It is important to immediately devise a time management strategy that works for you. Your goal by test day is to have a plan of action on how to work through the exam. As you are preparing for the exam, keep in mind that the clock will be working against you. Decide the maximum amount of time you will spend on one question before moving on to the next one. You do not want to run out of time.

WHAT TO BRING TO THE EXAM

Prior to test day, make sure you get a good night's rest and arrive armed with the following items:

A copy of the candidate information bulletin. Many times the proctor of the exam is inexperienced. If you are told the tabs are not allowed, you need to be able to defend yourself by referring to the candidate bulletin (the rulebook).

Bottle of water. The clock does not stop during the exam. There are no "hold" buttons. Should you need a sip of water and have to run to the water fountain or bathroom, your timer will continue to count down. Although you have little choice when it comes to a bathroom break, at least be prepared for the dry mouth syndrome.

Magnifying glass. Many of the documents/diagrams used throughout the exam are difficult to read. Save the frustration and headache-causing eyestrain and use a magnifying glass.

Two pencils and a pencil sharpener. Arrive with at least two pencils and a pencil sharpener, especially if you are taking the exam using the old fashioned pencil and paper format. Tests are still administered this way in several states and municipalities.

Two calculators. Remember Murphy's law, if it can go wrong it probably will on test day. Have a contingency plan for everything. If you insist on using your favorite calculator and it happens to be one that you are not sure is allowed, such as a construction master, make sure you have a backup.

A great attitude. Make every attempt to remain calm, cool, and collected. This is easier to maintain if you have adequate rest the night before the exam. Do not cram and stay up until midnight. This will work against your tolerance level for aggravation.

Remember, frustration will only create tension and make everything more difficult.

Make sure you also understand the requirements for identification, payment methods, and proper exam registration documentation.

LET THE TEST BEGIN

The moment you sit in the "hot seat" to begin the exam, let your strategy unfold. Have a plan and stick with it. A few recommendations:

1. **Switch to Test Mode.** Many times the rules of thumb and the assumptions you make in the field will not work in the test world. While your background and experience can be an asset, don't allow them to get in the way on test day. Make no assumptions. If you are not 100 percent sure about an answer, verify it at some point. Deliberately focus and concentrate on each question.

2. **Arrange your work area neatly.** Stack your books to one side and, if possible, stand them upright so they are each individually accessible. Decide that at least this one day you will be the most organized and careful person in the world.

3. **Download the memory.** That's right, download the information you are having trouble remembering. Transfer the formulas (and anything else you have been repeating since you walked into the test site) from your brain to the scratch paper provided by the proctor.

4. **Become familiar with the construction drawings and/or diagram booklet before you start the clock.** If your test is being administered by computer, you are in control of when the countdown begins. Take advantage of the control, but be careful not to push it. The proctor is only a few steps away and may prod you to begin the test if you wait too long.

5. **Answer the easy questions first.** Nothing will boost your confidence more than to run through a good portion of the exam answering questions based on information you recall from your studies. In contrast, your confidence level can diminish rapidly if you get distracted with a difficult question.

6. **"Mark" any question you answer and have doubt about.** If the test is computer based, you have the option to "mark" questions to review later. If you run out of time, the computer accepts the selected answer and does not penalize you for marking the question. By marking questions, if you have

Notes

additional time after answering all questions, you can verify the answers you selected.

7. **Leave the difficult questions unanswered and come back to these last.** As you skip questions, make a note on the scratch paper associating the question number with a specific book. This will allow you to categorize each of the unanswered questions by book saving valuable time and unnecessary frustration. After reading the last question on the exam, you will be allowed to revisit unanswered questions or marked questions. You can choose to go to the first of these in the group or to a specific question number.

8. **Do not run out of time.** Pay attention to the clock. Do not leave any question unanswered. Before you run out of time, select a choice for each question. Questions left unanswered will be counted against you. If you have to guess on a number of questions, improve your odds by selecting the same choice on each question. For example, select all As or all Cs but of course your goal is to manage your time and not have to guess on any of the questions.

The Dreaded Math

When it comes to the math portions of the exam, there are several very important and critical rules. It is in your best interest to convince yourself that for this one day, TEST DAY, you are going to take a methodical approach and deliberately follow each of the rules we are about to discuss.

1. **Always use a calculator.** This is probably the most critical and most important rule when it comes to math on the exam. Never perform mental calculations. Once you arrive at an answer, do the calculation again to check your answer. Repeat your calculations until you are certain the answer is correct.

2. **Use scratch paper to write down each step of math.** Professionals who use math on a daily basis, accountants, engineers, and the like, are sometimes the most difficult math students. Why? Because they refuse to write out the problem and their calculations, and they are reluctant to use a calculator for what they believe to be simple math. You do not want to fail this exam by one or two points; it hurts more to make a 69 than it does to make a 30. The point is, don't get in a hurry and miscalculate. Always write out your calculations and always calculate, at least twice, with a calculator.

3. **Never round until you get to the end.** This simply means that you should never round any of your numbers until you solve the equation. When necessary, round based on the choices or the directions provided in the question.

4. **Draw it out.** If you have to sketch apples and stick people, do it. Remember, you do not want to fail this test by one question. It is critical that you use your scratch paper to draw as you solve the equation to answer the question.

5. **Always multiply like numbers.** In other words, do not mix feet and inches. If you multiply 10 feet by 8 feet 6 inches, the answer is not 86, it is 85. You cannot simply multiply 10 by 8.6—you must first convert the 6 inches to feet.

6. **When converting inches to feet, round to the nearest hundredth of a foot.** Yes, this is an exception to rules. You round to the nearest hundredth when converting inches to feet because it puts your answer in the correct range of the choices provided and allows you to write it down more easily. You would not want to write 0.3333333333333; it would be much simpler to write 0.33.

Try these rules as you delve into the next section of the book, which covers the basics for performing necessary math calculations found on most plumbing exams. Good luck.

Notes

STATE LICENSING REQUIREMENTS FOR PLUMBING CONTRACTORS

State	Licensing Board Phone Number	Licensing Board Website	State Licensing Exam/ Testing Company	Pre-licensing or Pre-approval	Continuing Education
Alabama	(205) 945-4857	www.pgfb.state.al.us	Yes/PSI	Yes/Pre-approval	No
Alaska	(907) 465-3035	www.dced.state.ak.us/ occ/pcon.htm	Yes/Prometric	No	Yes
Arizona	(602) 542-1525	www.rc.state.az.us/	Yes/Prometric	No	No
Arkansas	(501) 661-2642	www.state.ar.us/clb	Yes/Prometric	Yes/Pre-approval	No
California	(800) 321-2752	www.cslb.ca.gov	Yes/California State Licensing Board	Yes/Pre-approval	No
Colorado	(303) 894-2300	www.dora.state.co.us/	Yes/Promissor	Yes/Pre-approval	No
Connecticut	(860) 713-6135	www.state.ct.us/dcp	Yes/PSI	Yes/Pre-approval	Yes
Delaware	(302) 744-4504	www.dpr.delaware.gov	Yes/Prometric	Yes/Pre-approval	No
Florida	(850) 487-1395	www.state.fl.us	Yes/Professional Testing Inc.	No	Yes
Georgia	(478) 207-1416	www.sos.state.ga.us/plb/ construct/	Yes/AMP	Yes/Pre-approval	Yes
Hawaii	(808) 586-2689	www.hawaii.gov/dcca/pvl	Yes/Prometric	Yes/Pre-approval	Yes
Idaho	(208) 334-3442	www2.state.id.us/	Yes/Plumbing Bureau	Yes/Pre-approval	Yes
Illinois	Not regulated by state/contact city/county		No		
Indiana	Not regulated by state/ contact city/county		No		
Iowa	(515) 281-7995	www.iowaworkforce.org/	No/Register with state; city/county may require testing		
Kansas	Not regulated by state/ contact city/county		No		
Kentucky	(502) 573-0364	www.ohbc.ky.gov	Yes/Plumbing Division	Yes/Pre-approval	No
Louisiana	(504) 826-2382	www.lslbc.state.la.us	Yes/State Plumbing Board	Yes/Pre-approval	No
Maine	(207) 624-8603	www.maineprofessionalreg.org	Yes/Prometric	Yes/Pre-approval	Yes
Maryland	(410) 230-6163	www.dllr.state.md.us	Yes/PSI	No	No
Massachusetts	(617) 727-9931	www.state.ma.us	Yes/PSI	Yes/Pre-approval	No
Michigan	(517) 241-9330	www.michigan.gov/dleg	Yes/State Plumbing Board	Yes/Pre-approval	No
Minnesota	(651) 284-5064	http://www.doli.state.mn.us/ plumbing.html	Yes/State Plumbing Board	Yes	No
Mississippi	(601) 354-6161	www.msboc.state.ms.us	Yes/PSI	Yes/Pre-approval	No

State	Phone	Website			
Missouri	Not regulated by state/contact city/county		No	No	
Montana	(406) 841-2367	www.discoveringmontana.com/dli/bsd/index.asp	Yes/State Plumbing Board	Yes/Pre-approval	Yes
Nebraska		No state licensing requirements at this time	No		
Nevada	(702) 486-1100	www.nscb.state.nv.us	Yes/PSI	Yes/Pre-approval	No
New Hampshire	(603) 271-3748	www.state.nh.us/plumbing/	Yes/International Code Council	No	No
New Jersey	(973) 504-6420	www.state.nj.us	Yes/State Plumbing Board	Yes/Pre-approval	Yes
New Mexico	(505) 452-8311	http://www.rld.state.nm.us/CID/index.htm	Yes/PSI	Yes/Pre-approval	No
New York	Not regulated by state/contact city/county		No		
North Carolina	(919) 875-3612	www.nclicensing.org	Yes/PSI	Yes/Pre-approval	No
North Dakota	(701) 328-9979	http://www.legis.nd.gov/information/rules/admincode.html	Yes/State Plumbing Board	Yes/Pre-approval	No
Ohio	(614) 644-3493	www.com.state.oh.us/dic/dicocilb.htm	Yes/International Code Council	Yes/Pre-approval	Yes
Oklahoma	(405) 271-5217	www.cib.sate.ok.us/	Yes/PSI	Yes/Pre-approval	Yes
Oregon	(503) 378-4133	www.oregonbcd.org	Yes/State Plumbing Board	Yes/Pre-approval	Yes
Pennsylvania	Not regulated by state/contact city/county		No/city/county may require testing		
Rhode Island	(401) 462-8000	www.dlt.state.ri.us	Yes/State Plumbing Board	Yes/Pre-approval	No
South Carolina	(803) 896-4686	www.llr.state.sc.us	Yes/PSI	No	No
South Dakota	(605) 773-3429	www.state.sd.us/dcr/plumbing	Yes/State Plumbing Board	Yes/Pre-approval	No
Tennessee	(615) 741-8307	www.state.tn.us/commerce/boards/contractors/	Yes/PSI	Yes/Pre-approval	No
Texas	(512) 458-2145	www.tsbpe.state.tx.us	Yes/State Plumbing Board	Yes/Pre-approval	Yes
Utah	(801) 530-6628	www.dopl.utah.gov	Yes/Prometric	Yes/Pre-approval	No
Virginia	(804) 367-8511	www.state.va.us/dpor	Yes/PSI	Yes/Pre-licensing	No
Washington	(360) 902-5207	www.lni.wa.gov	Yes/State Plumbing Board	Yes/Pre-approval	Yes
West Virginia	(304) 558-7890	www.labor.state.wv.us/	Yes/Prometric	No	No
Wisconsin	(608) 261-8500	www.commerce.wi.gov	Yes/State Licensing Board	Yes/Pre-approval	Yes
Wyoming	Not regulated by state/contact city/county		No		

PART ONE
Math

The following section covers the basic math concepts you are likely to encounter on your exam. It begins with a review of the terminology and simple concepts necessary to perform the calculations. Following the review, the 25 questions will test your general understanding of math as it applies to the plumbing industry.

Study Tip

When two negative numbers are multiplied together the product becomes positive.

NUMBERS

Whole Numbers

Whole numbers are positive numbers without any fractions or decimals.

Examples: $100.00; 8 hours; 12 wrenches

When we speak or write whole numbers, we automatically calculate certain numbers in our minds. . .

625: Six hundred plus twenty plus five
625: When we speak, we leave out the plus.
 Six hundred twenty-five

Mixed Numbers

Mixed numbers are whole numbers and parts of whole numbers. They can be expressed as fractions or as decimals.

Examples: $100.50; 7.5 hours; $5\frac{1}{3}$ inches

Decimal Numbers

A decimal number refers to part of a whole number. It is based on 10 or tenths broken down into hundredths, thousandths, etc., depending on how many numbers follow the decimal point.

Examples: 1.5 inches; 7.75 feet; 6.25 pounds

Fractional Numbers

A fractional number refers to part of a whole number expressed in a fraction instead of decimal form. It is possible to convert decimals to fractions and fractions to decimals.

Examples: $2\frac{2}{3}$ inches; $4\frac{1}{2}$" pipe; $1\frac{1}{4}$ miles

$$\frac{a}{b} = \frac{numerator}{denominator} = \frac{part\ of\ a\ whole\ number}{the\ whole\ number}$$

Fraction Bar

The fraction bar is the slash between the numerator and the denominator. It means to divide.

Example: $\frac{3}{4}$ means 3 divided by 4 or $3 \div 4$

Labeling Numbers

Labeling numbers is very important. On construction sites an estimator makes the calculations and architects draft the construction drawings. Labels are important to identify the unit of measurement. Label all calculations or lengths as inches, feet, yards, lengths of pipe, volume, pounds, etc.

Steps required for solving the equation:

1. Examine the problem and decide what information is given and what calculations are required.

2. Write down the information given and what requires an answer.

3. Decide which type of mathematical calculation is required to solve the problem.

4. Execute the complete operation.

5. Label the answer with the appropriate unit: inches, feet, BTUs, etc.

MATHEMATICAL SYMBOLS

/ Fraction bar

+ Add

− Subtract

× Multiply

÷ Divide

= Equals

≠ Does not equal

√ Square root

π Pi = 3.14

() Perform the calculation in the parentheses first

PIPING CALCULATIONS

Plumbers will at sometime or another calculate piping offsets. Offsets are pipe assemblies connecting a section of pipe from one line to another. The pipes are usually parallel. When using the right triangle theory, the length of the sides of a pipe offset can be calculated knowing the angle of the fittings and the length of one of the sides.

Study Tip

Mathematics should be completed in this order: parenthesis, exponents, multiplication, division, addition then subtraction.

Study Tip

The hypotenuse is commonly known as the travel.

A. Hypotenuse = Travel
B. Altitude = Advance
C. Base = Offset

Three important lengths include:

A. **Travel** (Hypotenuse) which is the distance from the initial offset to the second fitting.

B. **Advance** (Altitude) is the distance parallel to the pipes between the offset fittings.

C. **Offset** (Base) is the distance the pipe is offset from its original position to the new position.

Some of the pipe fitting angles that are used by plumbers for pipe offsets are:

- $22\frac{1}{2}°$ elbow or $\frac{1}{16}$ bend.

- 45° elbow or $\frac{1}{8}$ bend.

- 60° elbow or $\frac{1}{6}$ bend.

- 72° elbow or $\frac{1}{5}$ bend.

- 90° elbow or $\frac{1}{4}$ bend.

The plumber usually calculates the length of the travel. To calculate the travel, the angle of the fittings and the length of the advance or offset are required. The known values are multiplied by a constant. (See the single pipe offsets constants table below.) A 45° elbow ($\frac{1}{8}$ bend) is widely used by plumbers. The offset constant for this is 1.414 and should be memorized.

TABLE—SINGLE PIPE OFFSETS CONSTANTS							
Fitting	Fitting Angle	Travel = Offset ×	Travel = Advance ×	Advance = Offset ×	Advance = Travel ×	Offset = Advance ×	Offset = Travel ×
$\frac{1}{32}$	$11\frac{1}{4}°$	5.126	1.019	5.027	0.981	0.199	0.195
$\frac{1}{16}$	$22\frac{1}{2}°$	2.613	1.082	2.414	0.924	0.414	0.383
$\frac{1}{8}$	45°	1.414	1.414	1.000	0.707	1.000	0.707
$\frac{1}{6}$	60°	1.155	2.000	0.577	0.500	1.732	0.866
$\frac{1}{5}$	72°	1.051	3.236	0.325	0.309	3.078	0.951

Calculating Travel—Travel = Offset × Constant or Travel = Advance × Constant

Example:

Calculate the travel of a 45° elbow (⅛ bend) with an offset of 29".

1. Place the known values.

Travel = Offset × 1.414 (taken from offsets table on page 16)
Travel = 29" × 1.414 (offset is given value)

2. Calculate the travel.

29" × 1.414 = 41.006"
Travel = 41.006 approximately 41"

Calculate End-To-End

41 inches is the distance of the travel from center-to-center (C–C) measurement. Subtract twice the end-to-center calculating the end-to-end (E–E) measurement. This will be the length of pipe you should cut. The rule of thumb is that you subtract twice the size of the pipe being cut to allow for the fitting on each end.

Example:

In the above calculation the C–C measurement was 41 inches. If the pipe offset was 2 inches the E–E measurement would be 37 inches.

41" − 4" (2" for both fittings) = 37"

Multiple Pipe Offsets

Installing multiple pipes requires increased design ability by the plumber. Multiple pipes installed parallel to one another must be placed with a uniform spread. Spread is the center-to-center distance between the pipes. Offset elbows are installed evenly, guaranteeing an equal spread ensuring quality craftsmanship. 45° offsets are the best to use.

To install multiple pipes you must calculate the difference in the lengths of the pipes. The difference is added to or subtracted from the various pipe lengths used for the multiple pipe installation. Just like single pipe installations a constant value is used in the calculation.

Study Tip

Pythagorean theorem states that the square of the hypotenuse (side opposite of the 90° angle) of a right triangle is equal to the sum of the squares of the other two sides.

MULTIPLE PIPE OFFSETS Uniform Spread Constants S = Spread d = difference pipe length			
Fitting	Fitting Angle	Difference in length $d = S \times$	Travel = Offset ×
1/32	11¼°	0.099	5.126
1/16	22½°	0.199	2.613
⅛	45°	0.414	1.414
¼	60°	0.577	1.155
⅕	72°	0.727	1.051

Study Tip

Trigonometric functions are used to compute pipe offsets because of the dimensions known. Two functions most likely used are the sine and the tangent.

Calculating Multiple Pipe Offsets

An equation is used to calculate multiple pipe offsets. Two of the three values are given and you calculate the third. To calculate multiple pipe offsets, use the following equation:

$$d = S \times \text{constant}$$

The *d* value or difference equals the difference in length of the pipes. The value is obtained from the table on page 17 and is based on the fitting size.

Multiple Pipe Offsets

To calculate a multiple offset using a uniform spread:

1. Calculate the spread (*S*).
2. Multiply *S* by the constant. The constant value is given in the table Multiple Pipe Offsets and is based on the fitting size.
3. Calculate the *d* value.
4. Measure each offset pipe *d* value longer than the pipe installed next to it.

Example:

Calculate two pipes 12 inches center-to-center (C–C). The offset is 26 inches with 45° elbows. Calculate the length differences and the C–C length of the travel pipes.

$$d = S \times \text{constant}$$

1. Enter known values in the equation. 12 inches is given and the constant value is located on the Multiple Pipe Offsets illustration above.

$$d = 12" \times 0.414$$

2. Calculate the *d* value. This is the difference in offset pipe lengths.

$$d = 12" \times 0.414 = 4.968$$
4.968 is approximately 5"

Note: Although pipe lengths change, travel remains the same for all pipes.

Rolling Offsets

Sometimes a pipe changes direction both horizontally and vertically. This is called a rolling offset. The illustration below displays a rolling offset with 45° fittings and a 60° elbow. Often a wye fitting is used instead of the 60° elbow.

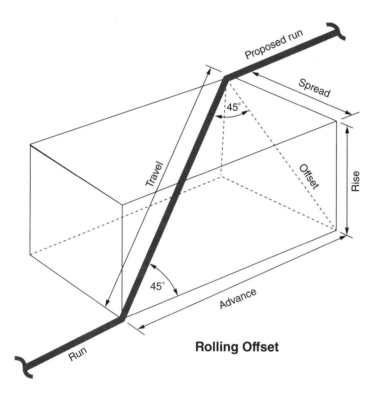

Rolling Offset

Squared values and square roots are used to calculate the offset distance. Squared values are written as 5^2 or whatever the number is $\#^2$. To calculate a squared value, take the given number and multiply the given number by itself.

$$2^2 = 4 \qquad 3^2 = 9 \qquad 4^2 = 16 \qquad 5^2 = 25$$

The offset is calculated by using the square root of the rise and spread. Rise is the actual vertical distance the pipe is offset and the spread is the horizontal distance the pipe is offset.

$$\text{Offset} = \sqrt{\text{rise}^2 + \text{spread}^2}$$

Study Tip

Rolling offset is the horizontal and vertical changes in direction of pipe.

Study Tip

There are 144 square inches in one square foot.

ROLLING OFFSETS CONSTANTS		
Fitting	**Travel = Angle**	**Travel = Offset ×**
1/32	11¼°	5.126
1/16	22½°	2.613
1/8	45°	1.414
1/4	60°	1.155
1/5	72°	1.051

$$offset = \sqrt{rise^2 + spread^2}$$

The square root of a number x is shown by \sqrt{x}. When multiplied by itself the square root produces the value x under the square root symbol. Use a calculator to calculate square roots quickly.

The square root of 4 is 2. $2 = \sqrt{4}$
The square root of 9 is 3. $3 = \sqrt{9}$
The square root of 25 is 5. $5 = \sqrt{25}$

A plumber must know the constant, the length of the rise, and the spread value when calculating the travel of a rolling offset.

To calculate a rolling offset:

1. Square the rise.
2. Square the spread.
3. Add the values of step 1 and step 2.
4. Calculate the square root of step 3.
5. Calculate your travel.

Example:

Calculate the travel for a 45° rolling offset with a rise of 5 inches and a spread of 13 inches.

$$Offset = \sqrt{rise^2 + spread^2}$$

1. Enter the known values in the equation: 5 inches is the given rise and the given spread is 12 inches.

$$Offset = \sqrt{5^2 + 13^2}$$

2. Calculate the squared values of the rise and spread.

$$Rise = 5^2 = 25$$
$$Spread = 13^2 = 169$$

3. Add the rise squared and the spread squared.

 $$25 + 169 = 194$$

4. Calculate the square root of 194.

 $$13.94 = \sqrt{194}$$

5. Place the known values to calculate the travel.

 Travel = Offset × 1.414 (given in Rolling Offsets Constants table)
 Travel = 13.94 (given in step 4) × 1.414

6. Calculate travel.

 Travel = 13.94 × 1.414 = 19.71
 Travel = 19.71" ≈ 19¾"

Common Rolling Offset

A common rolling offset used in plumbing is one that consists of a 45° wye and a 60° elbow. This is a very convenient set of values for a plumber. When these figures are used to implement a rolling offset, the rise and spread lengths are equal and the travel equals two times that length. For this exact situation, you can calculate the travel by multiplying the rise or the spread by 2.

 Situation = Rolling offset with a 45° wye and a 60° elbow
 Travel = Rise × 2 = Spread × 2

Pitch and Grade

Pitch and grade are used interchangeably. Drainage piping is set at a certain pitch (grade) to ensure adequate drainage. Pitch (P) is the slope of a line in respect to the horizontal plane. Pitch is displayed in inches per foot.

In the pipe's given length, fall (F) is quantity the line drops. Fall (drop) is expressed in inches.

Study Tip

Common rolling offset consists of a 45° wye and a 60° elbow.

Study Tip

Pitch is the degree of slope or grade given a horizontal run of pipe.

Pipe length (L) is the length of the pipe displayed in feet.

Several equations are possible when dealing with pitch.

Fall = Length × Pitch Pitch = Fall ÷ Length
Length = Fall ÷ Pitch Pitch = Length × % grade

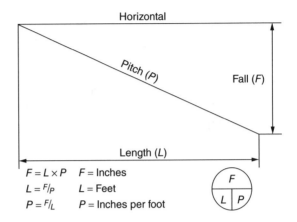

$F = L \times P$ F = Inches
$L = F/P$ L = Feet
$P = F/L$ P = Inches per foot

Examples:

Calculate the fall of a pipe run 32 feet long with a pitch at ¼" per ft.

Fall = Length × Pitch

1. Enter known values in the equation.

 Fall = 32' × ¼"

2. Calculate the value.

 Fall = ³²⁄₁ × ¼ = ³²⁄₄

3. Simplify the value.

 Fall = ³²⁄₄ = 8"

Calculate the length of a pipe with a pitch at ⅛ inch/foot and a fall of 40 inches.

Length = Fall ÷ Pitch

1. Enter the known values in the equation.

 Length = 40" ÷ ⅛

2. Calculate the value.

 Length = 40" × ⁸⁄₁ = ³²⁰⁄₁

3. Simplify the value.

 Length = ³²⁰⁄₁ = 320'

Calculate the pitch of a pipe that is 160 feet long and falls at 16".

Pitch = Fall ÷ Length

1. Enter known values in the equation.

 Pitch = 16" ÷ 160'

2. Calculate the value.

 ¹⁶⁄₁₆₀ = ¹⁄₁₀ inch/foot

Converting Percentage Grade to Fractional Inches per Foot

When converting a percentage grade to the equivalent fractional grade in inches per foot, divide the percentage grade by 100. This value is multiplied by 12 to calculate the decimal figure.

Pitch = (% grade ÷ 100) × 12

Example:

Calculate the fractional pitch for a 3% grade.

1. Enter the known values in the equation.

 Pitch = (3% ÷ 100) × 12
 Pitch = (3 ÷ 100) × 12

2. Calculate the value.

 Pitch = (³⁄₁₀₀) × 12
 Pitch = (0.03) × 12

Study Tip

Fall is the amount of slope given to a horizontal run of pipe.

Study Tip

A cubic foot of water equals 1728 cubic inches.

Pitch $= 0.03 \times 12 = 0.36$
Pitch $= 0.36 \approx$ ⅓ inch/foot
　　　3% grade $=$ ⅓ inch/foot

Converting Pitch in Inches per Foot to Percentage Grade

When converting a fractional grade in inches per foot to the equivalent percentage grade, convert the fractional pitch into a decimal and then divide by 12.

Example 1:

Calculate the percentage grade for the fractional pitch of ½" per foot.

1. Convert fraction into a decimal　　**½ = 1 ÷ 2 = 0.5**

2. Divide by 12　　**0.5 ÷ 12 = 0.04**

　　Percentage grade = **0.04 or 4%**

Example 2:

Calculate the percentage grade for the fractional pitch of ¼" per foot.

1. Convert the fraction to a decimal　　**¼" = 1 ÷ 4 = 0.25**

2. Divide by 12　　**0.25 ÷ 12 = 0.02**

　　Percentage Grade = **0.02 or 2%**

WATER CALCULATIONS

Water is present in three types of known states:

- Solid – Ice
- Liquid – Water
- Gas – Steam

When heat is added or taken away, a solid, liquid, or gas can change form. For instance, when heat is added to water, it will boil at 212°F changing the liquid to steam. When heat is taken away from water, it will freeze and become a solid at 32°F. Water, when frozen, expands to 8.33% of its normal volume, and it is the only substance on this planet where the maximum density of its mass does not occur when it becomes solidified.

For example, if a cubic foot of water (1728 cubic inches) freezes, you now have 1872 cubic inches of ice.

1728 + 8.33% = 1872 cubic inches

Calculating Water Pressure

Pressure is used to force water into and through buildings. Water pressure equals 0.434 pounds per square inch (psi) at the base of any container. Pressure gauges indicate the psi in the pipe to which it is connected. Most pressure gauges correct for the atmospheric pressure and are marked as pounds per square inch (psi). A head is a column of water. A 1-foot head equals a container with 1 foot of water. A 3-foot head contains three feet of water, and so on. Pipe is considered a container. For every 1-foot vertical increment, 0.434 psi is exerted. Diameter is not a factor. Pressure is directly related to the length of the pipe.

$$2.3 \text{ feet of head equals 1 psi of water}$$
$$2.3' = 1 \div 0.434$$

Given this calculation, for every foot of pipe, 0.434 psi is lost in pressure. This is a 1 psi loss for every 2.3 feet in height. To calculate the correct pressure or head for a building, you require the known head or the known pressure.

$$\text{Pressure} = \text{Head} \times 0.434$$
$$\text{Head} = \text{Pressure} \times 2.3$$

Study Tip

Psi is defined as pounds per square inch.

PRESSURE AND HEAD			
Head in Feet	**Pressure (psi) Pounds per Square Inch**	**Head in Feet**	**Pressure (psi) Pounds per Square Inch**
2.304	1	1	0.434
4.608	2	2	0.868
6.912	3	3	1.302
9.216	4	4	1.736
11.520	5	5	2.170
13.824	6	6	2.604
16.128	7	7	3.038
18.432	8	8	3.472
20.736	9	9	3.906
23.040	10	10	4.340

PSI and Heads Pressure = Head × 0.434 Head = Pressure × 2.3 Water = 62.5 pound per cubic foot

Study Tip

Pressure head is measured by the depth of a column of water, above the point where the measurement is taken.

Example:

Calculate the pressure created by a head 72 feet 6 inches.

Pressure = Head × 0.434

1. Enter the known values in the equation.

 Pressure = 72' 6" × 0.434
 Pressure = 72.5' × 0.434

2. Calculate the value.

 Pressure = 72.5' × 0.434 = 31.465
 Pressure ≈ 31½ psi

Example:

Calculate the head equivalency to a pressure of 60 psi.

Head = Pressure × 2.3

1. Enter the known values in the equation.

 Head = 60 psi × 2.3

2. Calculate the value.

 Head = 60 psi × 2.3 = 138 feet
 Head = 138 feet

Volume of a Pipe or Cylindrical Container

The volume of a pipe or cylindrical container can be calculated by multiplying $\pi \times r^2 \times h$ (3.14 × radius squared × height)

Example:

Calculate the volume of pipe 14 feet long with a radius of 2 inches.

Volume = $\pi \times r^2 \times h$

1. Enter the known values in the equation.

 Volume = $3.14 \times 2^2 \times 168$ inches (14' converted to inches)

2. Calculate the value.

 Volume = $3.14 \times 4 \times 168$
 Volume = 2110.08 cubic inches

Volume of a Pipe in Gallons

The volume, in gallons, for any length of pipe is calculated by multiplying 0.0408 by the inside diameter squared (D^2) by the length (L) in feet.

Volume = 0.0408 × D^2 × L

Example:

Calculate the volume, in gallons, for a pipe 10 feet long with a diameter of 6 inches.

1. Enter the known values in the equation.

 Volume (gal.) = $0.0408 \times 6^2 \times 10'$

2. Calculate the value.

 Volume = $0.0408 \times 36 \times 10'$

 Volume = 14.688 gallons

Note: This formula works because a cylinder 1 inch in diameter and 1 foot long holds 0.0408 gallons.

Weight of Water in a Pipe

When calculating the weight of water in a pipe, multiply 0.34 by the diameter squared (D^2) by the length (L) in feet of the pipe.

Weight = 0.34 × D^2 × L

Example:

Calculate the weight of water in a 6 inch pipe that is 8 feet long.

Study Tip

A gallon of water is equal to 8.33 pounds.

Study Tip

There are 231 inches of cubic water in 1 gallon.

1. Enter the known values in the equation.

$$\text{Weight} = 0.34 \times 6^2 \times 8'$$

2. Calculate the value.

$$\text{Weight} = 0.34 \times 36 \times 8'$$
$$\text{Weight} = 97.92 \text{ pounds}$$

Converting Cubic Inches to Gallons of Water

231 cubic inches of water equals 1 gallon of water. To calculate gallons divide the volume of water by 231.

Volume = Cubic Inches ÷ 231

Example:

Convert the volume of a container with 5775 cubic inches.

1. Enter the known values in the equation.

$$\text{Volume} = 5775 \div 231$$

2. Calculate the value.

$$\text{Volume} = 5775 \div 231 = 25 \text{ gallons}$$

PART TWO
Math Practice Exam

Study Tip

Markup determines the amount of profit to be made.

DETERMINING MARKUP

If a plumber purchases a water closet tank repair part for $15.00 and sells it to the customer with a 30½% markup, how much will the customer be charged?

1. Enter the known values in the equation.

 30½% = 30.5%

2. Convert the percentage to a decimal. Do this by moving the decimal point to the left two places.

 30.5% = 0.305

3. Multiply the cost of the repair part by 0.305

 $15 × 0.305 = $4.575 or $4.58

4. Add the markup price to the original cost.

 $15.00 Cost
 +4.58 Markup
 ————————————
 $19.58 Selling Price

Note: Insert the markup you desire in place of the 30½%

Percentage, Base, and Rate

You may be required to calculate a specific percentage (P), base (B), or rate (R).

$$P = B \times R$$
$$B = P \div R$$
$$R = P \div B$$

Example:

Figure the payment percentage due on a job. Initially, a $200 downpayment was made. The contract asks for $600 total.

1. Enter the known values in the equation.

$$P = B \times R$$
$$\$200 = \$600 \times R$$

2. Divide both sides by 600.

$$\text{}^{200}\!/\!_{600} = 1 \times R$$

3. Reduce the fraction.

$$\text{}^{1}\!/\!_{3} = R$$

4. Convert the fraction to a percentage.

$$\text{}^{1}\!/\!_{3} = 0.33$$
33% was paid up-front
67% is left to be paid

Study Tip

Laying length is the amount the pipe must be shortened to make the assembly of pipe and fittings equal to the theoretical length.

PRACTICE EXAM

1. What is an offset?

 a. The vertical distance between two pipes

 b. Pipe assemblies connecting a section of pipe from one line to another

 c. The length of pipe less the fittings

 d. All the above

2. Calculate the travel of a 45° offset with an offset of 28 inches.

 a. 39½"

 b. 33½"

 c. 11⅜"

 d. 37¼"

3. Calculate the travel of a 45° rolling offset with a rise of 8 inches and a spread of 20 inches.

 a. 30"

 b. 29¾"

 c. 30½"

 d. 33¼"

4. Calculate the fall of a drainage pipe 110 feet at a 4% grade.

 a. 4.4'

 b. 5'

 c. 4'

 d. 4.875'

5. Calculate the pressure created by a head 50 feet 9 inches.

 a. 20.23 psi

 b. 21 psi

 c. 24 psi

 d. 22.03 psi

6. What is the volume of a pipe 14 feet long with a diameter of 6 inches?

 a. 21.7836 gallons

 b. 20.5632 gallons

 c. 14.3268 gallons

 d. 19.8887 gallons

7. A customer wanted 40 psi at a building with only 20 psi available. The plumber was told to install a water tower. How much head would he need to have the 40 psi?

 a. 40'

 b. 17.36'

 c. 92'

 d. 86.5'

8. Fractional numbers _____.

 a. are part of the whole number

 b. are expressed in decimal form

 c. can be converted to decimals

 d. Both a and c

9. Which of the following is not considered when installing multiple pipe offsets?

 a. Spread

 b. Difference in pipe lengths

 c. Travel

 d. None of the above

Study Tip

PSF is pressure per square foot.

10. Whole numbers are _____.

 a. whole numbers and fractions

 b. positive numbers without fractions or decimals

 c. positive numbers with fractions and decimals

 d. numbers with decimals

11. Water, when frozen, expands _____ of its normal volume.

 a. 7.5%

 b. 8.33%

 c. 9.25%

 d. 9.76%

12. You are to install a new sewer line from building to sewer main. The sewer at the building is 3 feet deep. The ground surface is level. The distance from building to sewer main is 75 feet. How much deeper (fall) will the pipe need to be at the main sewer if the pitch (slope) is ⅛ inches per foot?

 a. 9⅜"

 b. 19½"

 c. 12¼"

 d. 14½"

 e. None of the above

13. The pressure at street level is 60 psi. You must run a water service to a house built on a 30 foot rise. What is going to be the pressure at the house?

 a. 60 psi

 b. 13.8 psi

 c. 34 psi

 d. None of the above

14. What is the one constant to memorize when calculating pipe offsets?

 a. 0.0408

 b. 1.414

 c. 2.3

 d. 0.434

15. When calculating pipe offsets, what must be given?

 a. Constant

 b. Offset (base)

 c. Advance (altitude)

 d. Either b or c

 e. All of the above

16. Calculate the weight of water in a 10-inch pipe that is 10 feet long.

 a. 340 lbs.

 b. 34 lbs.

 c. 340 psi

 d. 34 psi

17. Calculate volume of a tank (pipe) 20 feet in height with a diameter of 48 inches.

 a. 36,191.232 cu. in.

 b. 31,619.623 cu. in.

 c. 24,915.167 cu. in.

 d. 434,294.78 cu. in.

Study Tip

Hypotenuse is calculated as a constant of 1.414.

Study Tip

Volume equals $\pi r^2 h$.

18. How many gallons does the tank in question 17 hold?

 a. 1500 gal.

 b. 1880 gal.

 c. 2127 gal.

 d. 2475 gal.

19. Calculate a 1½ inch pipe offset that has an advance of 48 inches.

 a. 67⅞"

 b. 72⅞"

 c. 64⅞"

 d. 67⅞"

20. You buy a pressure reducing valve (PRV) for $35.98 to install for a customer. You want to make a 33⅓% markup. What would you charge the customer?

 a. $47.96

 b. $35.98

 c. $48.00

 d. $62.36

21. Why are labels on pipe, valves, etc., important?

 a. Labels identify hazards

 b. Labels identify contents

 c. Labels identify units of measurements

 d. All of the above

22. Your estimate totals $4,450.00 to replace a 4-inch sewer line. You want 60% down and 40% balance upon completion of job. How much do you need to collect before you start the job?

 a. $2,225.00

 b. $2,670.00

 c. $1,780.00

 d. None of the above

23. You estimated a job for $2,195.00. You agree to $900.00 up front with the balance paid when job is completed. What is the percentage you will collect up front?

 a. 33⅓%

 b. 45%

 c. 41%

 d. 38½%

24. What is the balance due at the completion of the job from question 23?

 a. $1,295.00

 b. $900.00

 c. Neither a nor b

 d. Both a and b

25. What does a *fraction bar* mean?

 a. Multiply

 b. Add

 c. Subtract

 d. Divide

Study Tip

Labor costs, material costs, and rental equipment are a few examples of what is included in an estimate.

Notes

PART THREE
Pipe, Fittings, and Symbols

This section serves as a review of pipe, fittings, and symbols used throughout most plumbing exams. A review of this section will prove beneficial in understanding the code, math, general knowledge, and identification practice exams that follow.

Study Tip

A drain is sized according to the drainage fixture load value assigned to each fixture.

PIPES

Drainage Pipe

Drainage pipes encountered by plumbers might be cast iron, steel, galvanized, DWV copper, schedule 40 PVC DWV plastic, clay, cement, or lead. The most common pipe used today is schedule 40 PVC because it is easy to work with and it is economical.

Cast Iron: Comes in single hub 5' and 10' and 5' double hubbed. Also comes in standard and extra-heavy weights. Color is black.

Steel: Usually comes in 20- or 21-foot lengths. Color is black.

Galvanized Steel: Usually comes in 21-foot lengths. Color is black.

DWV Copper: Usually comes in 20-foot lengths. Very expensive. Color is copper with a yellow stripe lengthwise.

PVC Schedule 40 DWV: Comes in various lengths. Color is white.

Plastic: A lightweight pipe used for leach field lines. Comes with or without holes for drainage. Comes in different colors, mostly white.

Clay: Used mostly in main sewer and storm drainage systems.

Cement: Used mostly in storm drainage systems.

Note: Lead pipe is now prohibited because of the potential hazard of lead poisoning.

Water Distribution Pipe

Water distribution pipes that you will encounter might be galvanized steel, copper types K, L, and M, schedule 40 PVC, CPVC, PE (polyethylene), PB (polybutylene), and PEX.

Copper Type K: Heaviest gauge used in commercial, industrial, and for underground water service lines. Color code is green.

Copper Type L: Most used of the copper types. Can be used for hot- or cold-water lines. Comes in various sizes ¼" to 4", also various lengths 2' to 20'. Color code is blue.

Copper Type M: Used for inside building and in walls. Not to be used in the ground, very thin. Comes in various sizes ½" to 2", and various lengths 5' to 20'. Color code is red.

PVC Schedule 40: Used for cold-water distribution only. Can be used for water service lines. Comes in various sizes ½" to 4", and comes in various lengths 5' to 20'. Color is white.

CPVC: Used for hot- and cold-water distribution. Comes in various sizes ½" to 2", and various lengths 5' to 20'. Color is cream.

PE (Polyethylene): Usually sold in coils. Used for water supply systems, jet wells, farm sprinkler systems, chemical well lines, and gas gathering systems. Joined together with plastic insert fittings and clamps. Sold in sizes ¾", 1", 1½", and 2". Color is black.

PB (Polybutylene): Flexible tubing (old style was gray—new style is milky white). Comes in 20-foot lengths and 50- to 1000-foot coils. Used for residential water supply only, mostly in mobile homes.

PEX: One of the most versatile, easy-to-use plumbing systems available. Can be used for hot- or cold-water systems. Uses brass fittings and copper crimp rings. A manifold can be used which has a shut-off valve for each line. Color usually is white. Can be color red for hot and color blue for cold.

Note: Be sure and check local codes. Some types of pipe may be prohibited in your area.

Study Tip

The longer the pipe, the more the pressure will drop since there is more friction to overcome.

Study Tip

Tees are fittings that create 90° branches in the piping.

FITTINGS

Elbows: Should be read as follows:

$\frac{1}{16}$ bend or 22½° bend
$\frac{1}{8}$ bend or 45° bend
$\frac{1}{4}$ bend or 90° bend
$\frac{1}{6}$ bend or 60° bend
$\frac{1}{5}$ bend or 72° bend

Tees and Wyes: Should be read by reading the flow first and then the side inlet. If it is a reducing fitting, the largest flow side should be read first.

Example:

A tee that has a flow of ¾" and a side inlet of ½" would be read: ¾ × ¾ × ½ tee, but when the flow sizes are the same, you read it: ¾ × ½ tee. The second ¾ is understood.

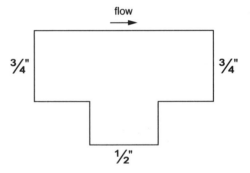

Example:

The same rules apply to reading drainage fittings:

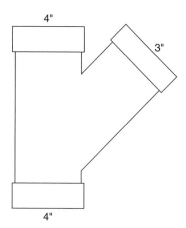

Read: 4 × 3 reducing wye

Drainage Fittings

Cast-iron fittings

| Reducing wye | Sanitary tee | Wye |

| Tee wye | Tapped tee | Tee cleanout |

Offset Reducer P-trap

Running trap Closet bend

PVC Schedule 40 DWV Fittings

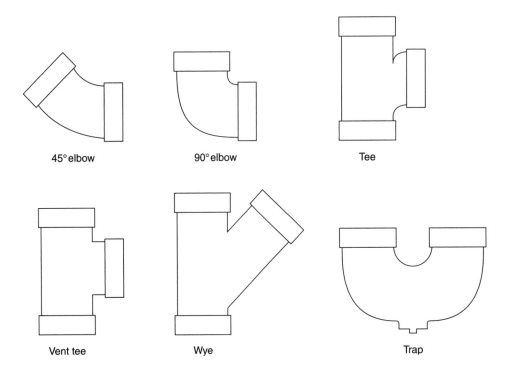

45° elbow 90° elbow Tee

Vent tee Wye Trap

WATER DISTRIBUTION FITTINGS

Galvanized Fittings

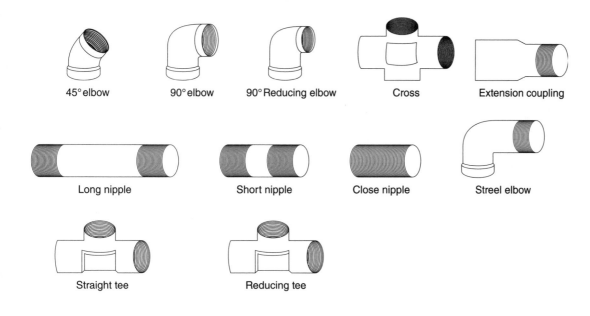

45° elbow 90° elbow 90° Reducing elbow Cross Extension coupling

Long nipple Short nipple Close nipple Streel elbow

Straight tee Reducing tee

Copper Fittings

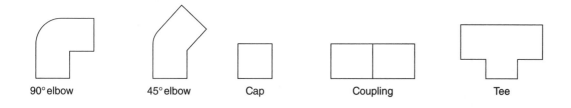

90° elbow 45° elbow Cap Coupling Tee

CPVC Fittings

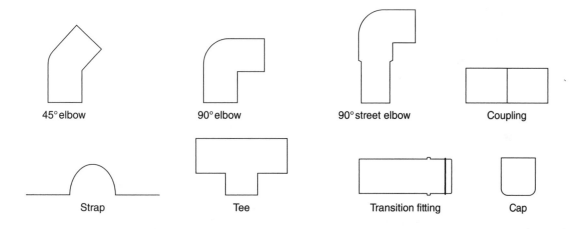

45° elbow 90° elbow 90° street elbow Coupling

Strap Tee Transition fitting Cap

PE (Polyethylene) Fittings

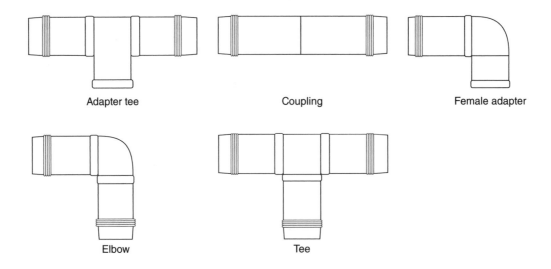

Adapter tee Coupling Female adapter

Elbow Tee

PB (Polybutylene) Fittings

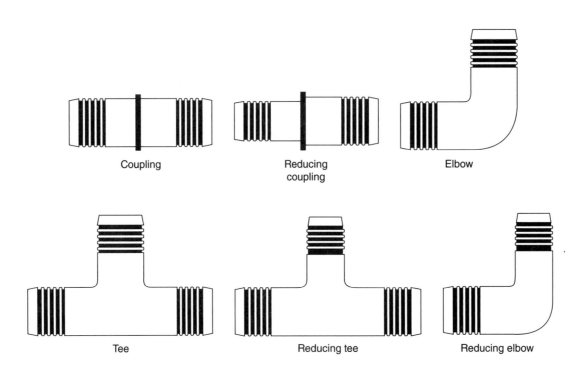

Coupling Reducing coupling Elbow

Tee Reducing tee Reducing elbow

PLUMBING/BLUEPRINT SYMBOLS

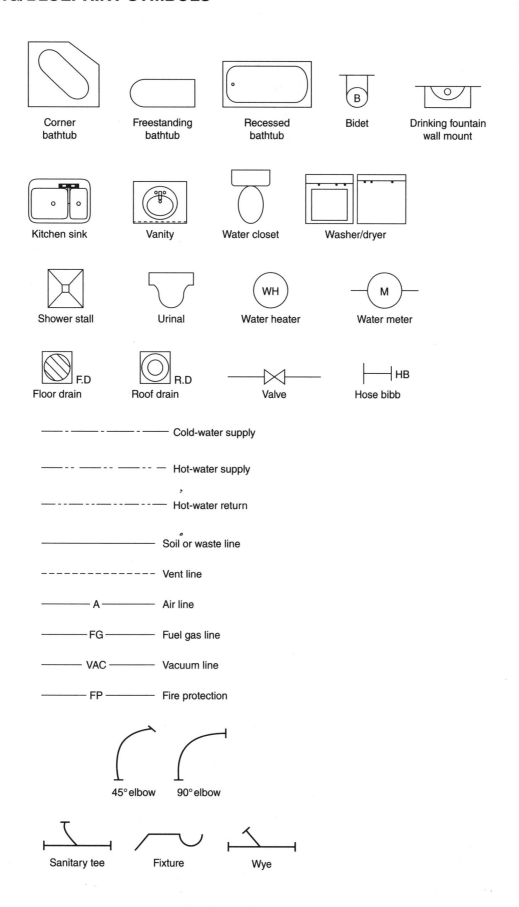

Corner bathtub

Freestanding bathtub

Recessed bathtub

Bidet

Drinking fountain wall mount

Kitchen sink

Vanity

Water closet

Washer/dryer

Shower stall

Urinal

Water heater

Water meter

F.D
Floor drain

R.D
Roof drain

Valve

HB
Hose bibb

Cold-water supply

Hot-water supply

Hot-water return

Soil or waste line

Vent line

A — Air line

FG — Fuel gas line

VAC — Vacuum line

FP — Fire protection

45° elbow

90° elbow

Sanitary tee

Fixture

Wye

Study Tip

PVC is used for cold-water piping and CPVC is used for hot-water piping.

PIPE, FITTINGS, AND SYMBOLS QUESTIONNAIRE

1. Which elbow(s) are read correctly?

 a. ¼ bend or 90° bend

 b. ⅛ bend or 45° bend

 c. ¹⁄₁₆ bend or 22½° bend

 d. All of the above

2. Identify this blueprint symbol.

 a. Vanity

 b. Water closet

 c. Bidet

 d. Toilet

3. Where can you use PVC Schedule 40 water pipe?

 a. All water distribution systems

 b. Cold-water distribution systems

 c. Hot-water distribution systems

 d. Hot-water heating systems

4. Where can you use CPVC water pipe and fittings?

 a. Liquid propane distribution systems

 b. Cold-water distribution systems

 c. Hot-water distribution systems

 d. Both b and c

5. Identify this blueprint symbol.

 a. Shower stall

 b. Washer

 c. Dryer

 d. Laundry sink

6. Identify this plumbing symbol.

 a. Tee

 b. Wye

 c. Valve

 d. Cross

7. Some of the pipes used for water distribution are

 a. CPVC

 b. Copper

 c. PEX

 d. PVC DWV

 e. a, b, and c

8. Drainage pipe/s you might encounter is/are

 a. CPVC

 b. PEX

 c. PE

 d. None of the above

9. What does this blueprint symbol represent?

 a. Air line

 b. Cold-water line

 c. Hot-water line

 d. Fire-protection line

10. Identify this fitting.

 a. Cleanout tee

 b. Sanitary tee

 c. Vent tee

 d. None of the above

Study Tip

To keep drawings to relatively small sizes, fractions are used to represent feet. This is known as scale drawing.

11. This fitting is used with what kind of pipe?

 a. PB (polybutylene)

 b. CPVC

 c. PVC

 d. Copper

12. What does this blueprint symbol represent?

 a. Cold-water line

 b. Hot-water return line

 c. Hot-water line

 d. Air line

13. What does this blueprint symbol represent? ___ FG ___ FG ___

 a. Fuel gas line

 b. Fuel gathering line

 c. Fire prevention line

 d. Drainage pipe line

14. Identify this fitting.

 a. Long turn elbow

 b. 90° bend

 c. Closet bend

 d. ¼ bend

15. Identify this fitting.

 a. P-trap

 b. Running trap

 c. S-trap

 d. Drum trap

16. What are the four types of copper water pipe?

 a. Water, drainage, gas, vent

 b. K, L, M, and DWV

 c. Gauge 8, 10, 12, and 14

 d. None of the above

Notes

17. What size tee is represented in this drawing?

 a. 1 × ¾ tee

 b. 1 × ¾ × ½ tee

 c. 1 × ½ × ¾ tee

 d. ¾ × ½ × 1 tee

18. Which water distribution pipe is one of the most versatile, uses brass fittings and copper crimp rings, and is the easiest to use for plumbing systems today?

 a. CPVC

 b. PE

 c. PVC

 d. PEX

19. What size fitting is represented in this drawing?

 a. 4" × 2" sanitary tee

 b. 4" × 2" wye

 c. 4" × 4" × 2 wye

 d. Either b or c

20. What can PE (polyethylene) pipe be used for?

 a. Water supply systems

 b. Jet wells

 c. Both a and b

 d. Drainage systems

Study Tip

Nipples are pieces of pipe that are threaded at both ends to join two fittings.

21. What is this fitting called?

 a. Street ell

 b. 90° elbow

 c. Both a and b

 d. 45° elbow

22. What is this fitting called?

 a. Short nipple

 b. Close nipple

 c. Short pipe

 d. All of the above

23. What is most common used drainage pipe today?

 a. Copper DWV

 b. Clay DWV

 c. Cement DWV

 d. PVC Schedule 40 DWV

24. Type M copper water pipe can be used _____.

 a. inside the building

 b. underground

 c. Both a and b

 d. None of the above

25. When installing water systems you should _____?

 a. check with local codes

 b. install the cheapest material available

 c. install good quality material

 d. Both a and c

PART FOUR
International Plumbing Code Practice Exams

The following section includes six practice exams based on the International Plumbing Code. You will also find several work sheets in this section outlining the steps involved to properly solve various equations necessary when using Chapter 7 of the *International Plumbing Code*. By completing each of the practice exams, you will become very familiar with the way the code book is formatted. Chapters 1 and 2 of the *International Plumbing Code* are not covered due to the fact that they cover administrative issues and definitions. However, it is recommended that you look through each of these chapters.

Study Tip

Tempered water shall be a maximum of 110°F.

IPC PRACTICE EXAM #1

The following questions are based on Chapters 3 and 4 of the *International Plumbing Code* and cover general regulations, fixtures, faucets, and fixture fittings.

1. Pipes passing through a concrete wall must be protected against corrosion. What is the minimum wall thickness of the protective sheathing?

 a. ⅛"

 b. ¼"

 c. ⅜"

 d. ½"

2. Exterior water supply system piping shall be installed not less than _____ inches below the frost line.

 a. 6

 b. 12

 c. 18

 d. 24

3. Sheet lead for pans shall not weigh less than _____ pounds per square foot.

 a. 1

 b. 2

 c. 3

 d. 4

4. CPVC pipe 2 inches in diameter must be horizontally supported every _____ feet.

 a. 3

 b. 4

 c. 10

 d. 15

5. Urinals shall NOT be substituted for more than ____ of the required water closets in assembly and educational occupancies.

 a. 25%

 b. 33%

 c. 50%

 d. 67%

6. What is the minimum size drain required for a floor drain?

 a. 1½"

 b. 2"

 c. 2½"

 d. 3"

7. Which of the following products or materials is required to be third party tested?

 a. Plumbing appliances

 b. Backflow prevention devices

 c. Plumbing fixtures

 d. Water distribution safety devices

8. A 10,000 sq. ft. museum has a maximum occupant load of 1,500 people. What is the minimum number of water closets required for female patrons?

 a. 8

 b. 10

 c. 12

 d. 23

9. Trenches installed parallel to footings shall not extend below the _____ bearing plane of the wall.

 a. 20°

 b. 45°

 c. 60°

 d. 90°

Study Tip

Urinals can be substituted for 67% of water closets.

Study Tip

Backflow is pressure created in the water distribution system.

10. A dishwasher shall be protected against backflow by a(n) _____.

 a. air break

 b. air gap

 c. vacuum breaker

 d. check valve

11. Joints at the roof and around vent pipes shall be properly flashed and made watertight by the use of ____ or other approved flashings or flashing materials.

 a. Lead

 b. Galvanized steel

 c. Aluminum

 d. All of the above

12. 1½ inch copper tube shall have a maximum horizontal hanger spacing of _____ feet.

 a. 6

 b. 10

 c. 12

 d. 15

13. Patrons of a mall shall not be required to travel more than ___ feet to access toilet facilities.

 a. 100

 b. 200

 c. 300

 d. 500

14. Shower compartments shall have a minimum cross-sectional area of _____.

 a. 900 sq. ft.

 b. 900 sq. in.

 c. 70 sq. in.

 d. 30 sq. in.

15. Coin-operated laundries shall be provided with floor drains having a minimum outlet of _____ inches.

 a. 2

 b. 3

 c. 4

 d. None of the above

16. What is the minimum slope required for the discharge of liquid combustion by products from condensing appliances?

 a. 1%

 b. 2%

 c. 3%

 d. 8%

17. Which of the following pipes is not required to be protected by shield plates?

 a. Copper pipe

 b. PVC

 c. Galvanized steel

 d. Brass pipe

18. How many water closets must be provided for a bar with a peak expected occupancy of 400?

 a. 5

 b. 8

 c. 10

 d. 12

Study Tip

Condensate piping shall be of approved corrosion-resistant material and not smaller than the drain connection to the appliance.

Study Tip

Water fountains are prohibited in restrooms.

19. 15 water fountains are required to be installed in a facility. What is the maximum number of bottled water dispensers allowed to substitute these fountains?

 a. 6

 b. 7

 c. 8

 d. 9

20. Trough urinals are _____.

 a. required

 b. permitted

 c. allowed only when approved by an official

 d. prohibited

21. Which of the following is used to support piping?

 a. Hangers

 b. Anchors

 c. Supports

 d. All of the above

22. Shower receptors shall be constructed of _____ materials.

 a. impervious and non-absorbent

 b. waterproof

 c. noncorrosive

 d. All of the above

23. A galvanized steel pipe that passes through a foundation wall requires protection from a sleeve that is _____.

 a. twice its outside dimension size

 b. two pipe sizes greater

 c. one pipe size greater

 d. More information is needed

24. Marble vanity tops with an integral lavatory shall conform to _____.

 a. CSA B45.1

 b. CSA B45.2

 c. CSA B45.4

 d. CSA B45.5

25. In a concealed location, piping shall be protected by a steel plate and shall extend a minimum of _____ inches above sole plate and below top plates.

 a. ½

 b. ¾

 c. 1

 d. 2

Study Tip

A strike plate prevents any physical damage to piping located in a concealed location.

Study Tip

Interconnections between hot- and cold-water supply piping systems shall be made to prevent flow between such piping systems.

IPC PRACTICE EXAM #2

The following questions are based on Chapters 5 and 6 from the *International Plumbing Code* and cover water heaters, water supply, and distribution.

1. Disregarding exceptions, all of the following are considered individual water supplies except _____.

 a. drilled wells

 b. dug wells

 c. streams

 d. land cisterns

2. What is the maximum distance a pan drain can terminate above the adjacent ground surface?

 a. 6"

 b. 12"

 c. 18"

 d. 24"

3. What is the minimum flow pressure of a temperature-controlled shower?

 a. 8 psi

 b. 15 psi

 c. 20 psi

 d. 25 psi

4. Water service pipe and the building sewer are required to be separated by a minimum of _____ feet.

 a. 5

 b. 8

 c. 10

 d. 12

5. The overflow for a water supply tank shall be covered by a corrosion-resistant screen of not less than _____ mesh per inch.

 a. 12 × 20

 b. 16 × 20

 c. 18 × 20

 d. 20 × 20

Study Tip

Two types of energy sources for water heaters are gas and electric.

6. Hospital fixtures shall be protected against backflow with all of the following except

 a. Reduced pressure principle backflow preventer

 b. Dual-check valve

 c. Spill-proof vacuum breaker

 d. Air gap

7. When tankless water heaters are intended for domestic use, the temperature of the water shall not exceed _____ degrees.

 a. 140

 b. 160

 c. 180

 d. 200

8. Water heaters operating above the atmospheric pressure shall be provided with _____.

 a. a levered pressure relief valve

 b. a temperature valve

 c. a self-closing pressure relief valve

 d. a levered pressure relief valve and temperature valve or combination thereof

Study Tip

Transition unions connect two different types of pipe.

9. Pumps installed in basements shall be mounted on block not less than _____ inches above the floor.

 a. 6

 b. 12

 c. 18

 d. 24

10. Joints used between copper and galvanized steel pipes shall be made with (a) _____.

 a. dielectric fitting

 b. only a mechanical fitting

 c. special approved adaptor

 d. None of the above

11. Valves used in a water supply system shall have a maximum of _____ lead content.

 a. 0.2%

 b. 2%

 c. 4%

 d. 8%

12. Which of the following backflow prevention devices may be used in a high hazard application?

 a. Double check backflow prevention assembly

 b. Backflow preventer with intermediate atmospheric vent

 c. Barometric loop

 d. Dual-check valve

13. Hazardous piping systems must be identified and repeated at intervals not to exceed _____ feet.

 a. 25

 b. 30

 c. 50

 d. 75

14. Attics containing water heaters shall be provided with a passage-way not less than _____.

 a. 30" × 30"

 b. 22" × 30"

 c. 20" × 30"

 d. 20" × 22"

15. Which of the following fixtures require a minimum ½" supply pipe?

 a. Bidet

 b. Drinking fountain

 c. Flushing rim sink

 d. Wall hydrant

16. Unfired hot water storage tanks shall be insulated to _____.

 a. R-12.5

 b. R-15

 c. R-20

 d. R-30

17. What is the minimum water service pipe?

 a. ½"

 b. ¾"

 c. 1"

 d. 2"

18. Which of the following joints is not permitted between poly-ethylene plastic pipe and tubing?

 a. Flared joints

 b. Threaded joints

 c. Heat fusion joints

 d. Mechanical joints

Study Tip

All appliances must be provided with access.

Study Tip

A sill cock is an outside faucet used for connection to a garden hose.

19. Shutoff valves are required to be installed in all of the following locations except for _____.

 a. kitchen sinks in a single-family dwelling

 b. bathtubs in single-family dwellings

 c. water supply pipe to each sill cock

 d. water supply pipe to each appliance

20. What is the maximum flow pressure for a private lavatory?

 a. 2 psi

 b. 8 psi

 c. 15 psi

 d. 60 psi

21. Which of the following water service pipes must conform to ASTM D 2672?

 a. Polybutylene plastic pipe

 b. Ductile iron water pipe

 c. Polyvinyl chloride plastic pipe

 d. Chlorinated polyvinyl chloride plastic pipe

22. The required drain size for a water tank that has a capacity of 1800 gallons is _____.

 a. 1"

 b. 1½"

 c. 2"

 d. 2½"

23. What size overflow pipe is required for a water supply tank if the supply line to the tank has a capacity of 150–200 gpm?

 a. 2"

 b. 2½"

 c. 3"

 d. 4"

24. Cross connections between a private water supply and public supply shall be _____.

 a. permitted

 b. required

 c. prohibited

 d. None of the above

25. A 3-inch drain pipe can empty a water tank with a capacity of _____ gallons.

 a. 6,000

 b. 4,000

 c. 2,000

 d. 1,000

Study Tip

A well is considered a private water supply.

Study Tip

A cleanout plug is placed in drain pipe lines to afford access to the pipes for the purpose of cleaning their interior.

IPC PRACTICE EXAM #3

The following questions are based on Chapter 7 of the *International Plumbing Code* and cover sanitary drainage.

1. What is the maximum temperature wastewater can be when discharged into the building drainage system?

 a. 60°F

 b. 140°F

 c. 180°F

 d. 200°F

2. Brass cleanout plugs shall be used with _____.

 a. metallic drains

 b. waste piping

 c. vent piping

 d. All of the above

3. A vertical pipe to which connections are made from the fixture side of a trap and through which vapor is removed from a device utilized on bedpan washers is known as _____.

 a. stack vent

 b. vent stack

 c. individual vent

 d. local vent stack

4. A horizontal stack offset is located 5 branch intervals below the top of the stack. A horizontal branch interval shall not connect within _____ feet above or below such offset.

 a. 1

 b. 2

 c. 3

 d. 4

5. What is the minimum size sterilizer vent stack serving an instrument washer?

 a. 1"

 b. 1½"

 c. 2"

 d. 2½"

6. What size pipe is required for a building drain on ¹⁄₁₆" slope accepting discharge from 2,700 fixture units?

 a. 6"

 b. 8"

 c. 10"

 d. 12"

7. Joints between a cast iron and brass pipe shall be made with all of the following except _____.

 a. welded joint

 b. threaded joint

 c. caulked joint

 d. approved adapter fitting

8. A quarter bend is used as a change of direction from vertical to horizontal. What is the maximum size fitting allowed?

 a. 1½"

 b. 2"

 c. 3"

 d. 4"

9. What is the minimum slope for a 5-inch horizontal drain pipe?

 a. ½"

 b. ¼"

 c. ⅛"

 d. ¹⁄₁₆"

Study Tip

Federal law prohibits the use of solders containing over 0.2% lead in potable water systems.

10. Which of the following types of pipe shall not be used for underground building drainage?

 a. Cast-iron pipe

 b. Brass pipe

 c. DWV

 d. ABS

11. Caulking ferrules for sanitary drainage shall be of _____.

 a. red brass

 b. stainless steel

 c. plastic

 d. borosilicate glass

12. Where more than one change of direction occurs in a run of piping, only one cleanout shall be required for each ___ of developed length of the drainage piping.

 a. 40'

 b. 50'

 c. 75'

 d. 100'

13. A _____ located on the discharge side of the check valve shall be installed in the pump or ejector discharge piping between the pump or ejector and the gravity drainage system.

 a. check valve

 b. full open valve

 c. Both a and b

 d. Neither a nor b

14. A 4-inch horizontal drain can accommodate a maximum of _____ fixture units.

 a. 160

 b. 170

 c. 360

 d. 400

Study Tip

A sewage ejector lifts sewage by entraining the sewage in a high-velocity jet of steam, air, or water.

15. What is the minimum capacity of a sewage ejector with a 2½" discharge pipe?

 a. 21 gpm

 b. 30 gpm

 c. 46 gpm

 d. 50 gpm

16. A sump pit shall have a diameter at least _____ inches.

 a. 18

 b. 24

 c. 36

 d. 48

17. Solvent cement joints shall be permitted _____.

 a. above ground

 b. below ground

 c. both above and below ground

 d. None of the above

Study Tip

A bidet requires connecting the bowl outlet to the DWV piping system through a P-trap.

18. What is the drainage fixture unit value for a bidet?

 a. ½

 b. 1

 c. 2

 d. 3

19. Building sewers with manholes shall be located not more than _____ feet apart from the junction of the building sewer.

 a. 100

 b. 200

 c. 300

 d. 400

20. What is the fixture unit value for an unlisted fixture with a 3" trap?

 a. 2

 b. 3

 c. 4

 d. 5

21. What is the minimum size cleanout required for a 3" drain pipe?

 a. 1"

 b. 2"

 c. 3"

 d. 4"

22. A fixture having a continuous flow of 3 gpm is equivalent to how many fixture units?

 a. 3

 b. 4

 c. 5

 d. 6

23. What is the minimum trap size for a water fountain?

 a. 1¼"

 b. 1½"

 c. 2"

 d. 2½"

24. What is the minimum clearance for rodding a cleanout of a 4-inch pipe?

 a. 18"

 b. 24"

 c. 30"

 d. 36"

25. What is the total fixture unit value for 6 floor drains and 3 public water closets rated 1.6 gpf?

 a. 21 dfu

 b. 24 dfu

 c. 30 dfu

 d. 36 dfu

Study Tip

Water fountain rough-ins are correctly done when the supply and drain pipes are inside the enclosure of the unit.

Study Tip

Every type of fixture is assigned a fixture unit value.

Drainage Fixture Units

The following information is based on Chapter 7 of the *International Plumbing Code* and cover calculations used for computing drainage fixture unit values, horizontal drains, and building drains.

Drainage Fixture: A measurement of probable discharge by various types of plumbing fixtures into a drainage system.

To properly size horizontal drains, building drains, and vent systems, a total dfu value must be established for each of these.

IPC 2009, Table 709.1: This table displays the dfu value and minimum trap size for each type of fixture listed.

Example: What is the total dfu value for 2 (two) floor drains?

 a. 1
 b. 2
 c. 3
 d. 4 **Ans: d**

2 floor drains × 2 dfu = 4

Example: What is the total dfu value for 3 (three) automatic clothes washers in a self-service laundry?

 a. 3
 b. 6
 c. 9
 d. 12 **Ans: c**

3 clothes washers × 3 dfu = 9

IPC 2009, Table 709.2: In some circumstances, a specific fixture may not be included in Table 709.1. If this occurs, a dfu value can still be determined by the trap or drain size of the fixture. Table 709.2 assigns a dfu value for unlisted fixtures when either the drain or trap size is known.

Example: What is the drainage fixture unit value for a dishwasher in a restaurant on a 2-inch drain?

 a. 2
 b. 3
 c. 4
 d. 5 **Ans: b**

Building Drains

Building Drain: The lowest part of a drainage system that receives discharge from soil, waste, and other drainage pipes inside and delivers the waste to the building sewer.

3 Types of Building Drains

<u>Sanitary:</u> transports sewage only
<u>Storm:</u> transports storm drainage
<u>Combined:</u> transports storm drainage and sewage

A building drain is sized according to the tallied drainage fixture unit value conveyed through the drainage system.

<u>IPC 2009, Table 710.1(1):</u> This table limits the number of dfus connected to any portion of the building drain or the building sewer according to the slope per foot.

Example: A building drain having ¼" slope receives a total discharge of 200 dfus. What size drain is required to convey this amount of waste?

 a. 2½"

 b. 3"

 c. 4"

 d. 5" **Ans: c**

Example: A 5" building drain having a ½" slope is capable of conveying a maximum discharge of how many dfus?

 a. 480

 b. 575

 c. 840

 d. 1,000 **Ans: b**

Study Tip

Storm drainage piping system of a property is separate from the DWV system.

Study Tip

A branch interval is the vertical distance between the connections of pipes on the main DWV stack.

Horizontal Fixture Branches and Stacks

Horizontal Branch Drain: Drainage pipe extending laterally from a soil or waste stack or building drain that receives discharge from two or more fixture drains and transports the discharge to the soil, waste stack, or building drain.

Stack: vertical line of soil, waste, vent or inside conductor piping that extends through at least one story with or without offsets.

<u>**IPC 2009, Table 710.1(2):**</u> This table properly sizes horizontal drains and stacks according to the number of dfus being transported through them.

Column 1: Diameter of pipe

Column 2: Lists the maximum number of fixture units that may discharge into any size horizontal drain

Column 3:
Column 4: } sizes the stack according to the number of branch intervals
Column 5:

Example: What is the maximum number of fixture units allowed to discharge in a 4-inch horizontal branch?

 a. 90

 b. 160

 c. 200

 d. 360 **Ans: b (Columns 1 & 2)**

Example: A horizontal branch that receives 500 dfu of discharge is required to be a minimum of _____ inches.

 a. 4

 b. 5

 c. 6

 d. 8 **Ans: c (Columns 1 & 2)**

Example: What is the maximum number of fixture units allowed to discharge in a 5-inch stack with 2 branch intervals?

 a. 200

 b. 240

 c. 500

 d. 540 **Ans: d (Columns 1 & 4)**

IPC PRACTICE EXAM #4

The following questions are based on Chapter 7 of the *International Plumbing Code* and cover drainage fixture unit values, horizontal, drains, and building drains.

1. A water closet flushing 1.6 gpm and a lavatory are connected to a horizontal branch in a dwelling unit. What is the fixture unit total that will be discharging into this branch?

 a. 4 dfu

 b. 5 dfu

 c. 6 dfu

 d. 8 dfu

2. What size horizontal branch is required to convey the discharge to the building drain from question #1?

 a. 1½"

 b. 2"

 c. 2½"

 d. 3"

3. A clothes washer, two compartment laundry tray, and a floor drain are connected to a horizontal branch in a home. What size pipe diameter is required for drainage?

 a. 1½"

 b. 2"

 c. 2½"

 d. 3"

Study Tip

The stub-out must be cut off and a valve installed before the water closet is put in position.

Study Tip

A bathroom group consists of a water closet, lavatory, bathtub or shower, including or excluding a bidet, an emergency floor drain or both.

4. A four-story motel contains 6 sleeping rooms on each floor. All sleeping rooms include a bathroom group with a water closet flushing greater than 1.6 gpm. What is the total dfu value discharging into each branch interval?

 a. 30

 b. 36

 c. 120

 d. 144

5. An 8-inch stack is conveying 2,000 fixture units worth of drainage into the building drain. The building has five floors. How many dfus are emptying into this stack at each floor?

 a. 400

 b. 2,000

 c. 2,200

 d. 3,600

6. A ⅛-inch sloped building drain receives discharge from 2 branch intervals. Each branch interval conveys 120 fixture units worth of waste. What size building drain is required to transport the drainage to the sewer?

 a. 3"

 b. 4"

 c. 5"

 d. 6"

7. What is the maximum number of fixture units allowed to convey through a 6-inch building drain on a ½" slope?

 a. 575 dfu

 b. 840 dfu

 c. 1,000 dfu

 d. 2,300 dfu

8. Given: Horizontal branch #1 = 46 dfu
 Horizontal branch #2 = 38 dfu
 Horizontal branch #3 = 54 dfu

 Building drain's slope = ¼"

 What size building drain is required to convey the discharge from the horizontal branches listed above?

 a. 3"

 b. 4"

 c. 5"

 d. 6"

Study Tip

A horizontal branch extends laterally from a soil or waste stack or building drain.

Study Tip

At any point in the waste piping where there is a directional change, a relief vent will help eliminate backpressure buildup.

IPC PRACTICE EXAM #5

The following questions are based on Chapters 8 and 9 of the *International Plumbing Code* and cover indirect waste and vents.

1. All horizontal indirect waste pipes exceeding _____ feet must be trapped.

 a. 2

 b. 4

 c. 6

 d. 8

2. Standpipes shall extend a maximum of _____ inches above the trap weir.

 a. 12

 b. 18

 c. 36

 d. 42

3. A vent stack is required for every drainage stack that exceeds _____.

 a. 2 branch intervals

 b. 3 branch intervals

 c. 4 branch intervals

 d. 5 branch intervals

4. The developed length of a relief vent is measured from the farthest point of vent connection to the drainage system, then, to the point of connection to the _____.

 a. vent stack

 b. stack vent

 c. termination outside a building

 d. All of the above

5. A sump pump, other than pneumatic, has a discharge capacity of 60 gpm. What diameter of vent pipe is required if the developed length is 78 feet?

 a. 1¼"

 b. 1½"

 c. 2"

 d. 2½"

6. Where fixture drains connect at different levels and share a common vent, which fixture is not permitted at the uppermost level?

 a. Water closet

 b. Urinal

 c. Three-compartment sink

 d. Whirlpool spa

7. Air admittance valves are required to be installed how far above insulation materials?

 a. 6"

 b. 12"

 c. 18"

 d. 24"

8. A dry vent shall rise to a minimum of _____ above the flood level rim of the highest trap.

 a. 4"

 b. 6"

 c. 8"

 d. 10"

Study Tip

An air admittance valve allows air to enter the plumbing drainage system when negative pressure develops in the piping system.

Study Tip

Flashing prevents water from leaking into the building around the stack.

9. In order for any harmful chemicals to be discharged into a plumbing system, what must first happen to prevent injury to the drain or sewer?

 a. Chemicals must be diluted

 b. Chemicals must be neutralized

 c. Both a and b

 d. None of the above

10. A trap seal of any fixture shall not be subjected to a pneumatic pressure differential of more than _____.

 a. 0.5" WC

 b. 1" WC

 c. 2" WC

 d. None of the above

11. A vent that connects to a horizontal drainage branch and vents a minimum of two traps up to a maximum of eight is called a _____.

 a. common vent

 b. circuit vent

 c. branch vent

 d. individual vent

12. What is the maximum developed length of a 2½" vent?

 a. 230'

 b. 270'

 c. 300'

 d. 360'

13. An individual vent serving as a common vent is permitted to vent _____ traps.

 a. 2

 b. 3

 c. 4

 d. 6

Study Tip

A common vent connects at the junction of two fixture drains.

14. A 20-foot branch vent is 2 inches in diameter. What size branch vent would be required if the developed length is increased to 50 feet?

 a. Same size is required

 b. Increase the diameter by one nominal size for entire developed length

 c. Increase the diameter by two nominal sizes for entire developed length

 d. Increase the diameter by two and one-half sizes for entire developed length

15. What is the maximum distance from the trap weir to the vent if the drain size is 1¼"?

 a. 3½'

 b. 5'

 c. 6'

 d. 10'

16. Waste receptors shall be permitted in the form of a hub or pipe extending not less than _____ above a water-impervious floor and are not required to have a strainer.

 a. 1"

 b. 1½"

 c. 2"

 d. 2¼"

17. What is the minimum size wet vent required if the fixture unit discharge to the wet vent is 5 dfus?

 a. 1½"

 b. 2"

 c. 2½"

 d. 3"

18. What size relief vent is required if connected to a 4-inch vent stack?

 a. 1"

 b. 2"

 c. 3"

 d. None of the above

19. An air admittance valve is used to vent fixtures _____.

 a. once every 2 floors

 b. once every 3 floors

 c. on the same floor level

 d. None of the above

20. What is the maximum wet vent size permitted?

 a. 1½"

 b. 2"

 c. 2½"

 d. 3"

21. What is the minimum size air gap required for a 1½" indirect waste pipe?

 a. 1½"

 b. 2"

 c. 2½"

 d. 3"

22. Vent terminals extending through a wall shall terminate a minimum of _____ feet from a lot line.

 a. 10

 b. 15

 c. 20

 d. 25

Study Tip

Vent terminals shall not be used as a flag pole or television aerials.

23. Sheet copper for vent flashings must weigh a minimum of _____ ounces per square foot.

 a. 4

 b. 6

 c. 8

 d. 10

24. What is the maximum distance allowed from a fixture trap to the vent if the trap size is 1½" on ¼" slope?

 a. 3½'

 b. 5'

 c. 6'

 d. All of the above

25. What is the minimum size stack vent allowed to serve a drain?

 a. 1¼"

 b. 1½"

 c. 2"

 d. 2½"

Study Tip

A leader is an exterior drain pipe for conveying storm water from roof or gutter drains.

IPC PRACTICE EXAM #6

The following questions are based on Chapters 10 and 11 of the *International Plumbing Code* and cover traps, interceptors, separators, and storm drainage.

1. What is the maximum size liquid seal required for fixture traps?

 a. 1½"

 b. 2"

 c. 3"

 d. 4"

2. A cast-iron subsoil drain pipe must conform to all of the following standards except _____.

 a. ASTM B 74

 b. ASTM A 888

 c. CISPI 301

 d. ASTM A 74

3. What is the minimum slope required for the horizontal branches of a building storm drain?

 a. ⅛ unit vertical in 12 units horizontal

 b. ¼ unit vertical in 12 units horizontal

 c. ½ unit vertical in 12 units

 d. ¾ unit vertical in 12 units horizontal

4. A 12,000 square foot roof has six leaders installed for storm drainage. How many square feet of roof area is each leader responsible for draining?

 a. 6,000 sq. ft.

 b. 4,000 sq. ft.

 c. 2,000 sq. ft.

 d. Not enough information

5. A sump pit shall be constructed from what type of material?

 a. Plastic

 b. Steel

 c. Tile

 d. All of the above

6. Discharge piping from an ejector in a dwelling unit is required to have what kind of valve?

 a. Check valve

 b. Gate valve

 c. Globe valve

 d. No valve is required

7. Which of the following is a prohibited trap?

 a. Bell trap

 b. Crown-vented trap

 c. Drum trap

 d. All of the above

8. Which of the following is permitted to discharge storm water out onto a street?

 a. A drug store having at least 25' of clearance in front of building

 b. A ten-unit apartment complex

 c. A convenience store surrounded by a concrete parking lot

 d. A two-family dwelling

Study Tip

A drum trap is a vertically oriented cylinder with an inlet near the bottom outlet near the top.

Study Tip

*Grease interceptors shall
receive waste only from fix-
tures and equipment that
allows fats, oils, or grease to
be discharged.*

9. <u>Given:</u> 1 inch per hour rainfall rate

 Two gpm of continuous discharge into building storm drain is equivalent to _____ of roof area.

 a. 96 sq. ft.

 b. 192 sq. ft.

 c. 288 sq. ft.

 d. 384 sq. ft.

10. Roof drains have strainers which must extend a minimum of _____ inches above the roof's surface.

 a. 1½

 b. 2

 c. 3

 d. 4

11. What is the maximum distance allowed from a fixture outlet to a trap weir?

 a. 12"

 b. 18"

 c. 24"

 d. 36"

12. What is the retention capacity of a grease trap with a 7-gpm flow through rating?

 a. 14 lbs.

 b. 18 lbs.

 c. 20 lbs.

 d. 24 lbs.

13. <u>Given:</u> 7 inches rainfall per hour

 What is the maximum square feet of roof area a single 4-inch leader can convey?

 a. 1,260 sq. ft.

 b. 2,630 sq. ft.

 c. 3,070 sq. ft.

 d. 4,945 sq. ft.

14. What is the minimum diameter of a subsoil drain?

 a. 2"

 b. 3"

 c. 4"

 d. 6"

15. A sump pit's discharge piping must include a _____.

 a. full flow check valve

 b. double check valve

 c. gate valve

 d. gate valve and full flow check valve

16. What is the minimum number of roof drains required to be installed on a 10,000 square foot roof area?

 a. 2

 b. 4

 c. 5

 d. 10

Study Tip

Gate valves are best suited for main supply lines and pump lines.

Study Tip

An oil separator is not required in hydraulic elevator pits where an approved alarm system is installed.

17. Subsoil drains shall discharge to a _____.

 a. sump

 b. dry well

 c. approved location above ground

 d. All of the above

18. Which of the following is not required to have an oil separator?

 a. A mechanic's garage

 b. Commercial parking garage storing automobiles

 c. A parking garage which dispenses gasoline

 d. Car wash

19. Slip joints shall be installed only _____.

 a. on the trap inlet

 b. on the trap outlet

 c. within the trap seal

 d. All of the above

20. Sand interceptors shall have a water seal not less than _____.

 a. 6"

 b. 12"

 c. 18"

 d. 24"

21. What size trap is required for a 4-inch storm drain connected to a combined sewer?

 a. 3"

 b. 4"

 c. 5"

 d. None of the above

22. What is the minimum size water seal required of an outlet opening to an oil separator?

 a. 6"

 b. 12"

 c. 18"

 d. 24"

23. Given: 4" rainfall per hour
 ⅛" slope

 What size horizontal storm drain is required for 5,500 sq. ft. of roof drainage?

 a. 4"

 b. 5"

 c. 6"

 d. 8"

24. A 15,000 square foot roof area is required to have at least how many roof drains?

 a. 2

 b. 4

 c. 5

 d. 10

25. Given: 5" rainfall per hour
 4% slope

 What is the maximum square footage a 3-inch horizontal storm drain can convey?

 a. 657 sq. ft.

 b. 928 sq. ft.

 c. 1,310 sq. ft.

 d. 3,010 sq. ft.

Study Tip

In sizing roof drains and storm drainage piping, one-half of the area of any vertical wall that diverts rainwater to the roof shall be added to the projected roof area.

Study Tip

A typical storm drainage system consists of gutters, downspouts, and leaders.

STORM DRAIN SIZING

This section primarily addresses Tables 1106.2 and 1106.3 in Chapter 11 of the *International Plumbing Code*.

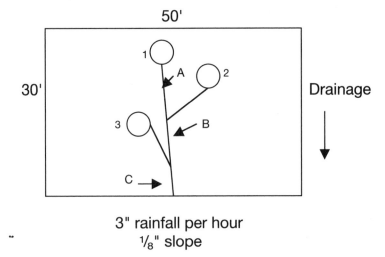

3" rainfall per hour
$\frac{1}{8}$" slope

○ = leaders
____ = horizontal storm drain piping

<u>Leader:</u> An exterior drainage pipe for conveying storm water from roof or gutter drains to an approved means of disposal.

I. *Sizing the Leaders: Table 1106.2*

1. Find the square footage of the roof
 30' × 50' = **1500 sq. ft.**

2. Divide the square footage of the roof by the number of leaders displayed in the above diagram.
 1500 sq. ft. ÷ 3 leaders = **500**

This means that each leader is responsible for conveying 500 sq. ft. worth of storm water drainage.

Using *Table 1106.2*, properly size the leaders according to code.

Rainfall rate: 3" per hour (find this rate at the top of the table)
The numerical values below this column are the projected square footage of drainage. We have already predetermined that each leader will be conveying 500 sq. ft. of storm water.

Find the numerical value that will accommodate at least 500 sq. ft. Once identified, follow that row to the left side of Table 1106.2 to determine the diameter of each leader.

<u>**Answer:**</u> **2"**

II. *Sizing Horizontal Storm Drain Piping: Table 1106.3*

Section A:

1. When sizing horizontal storm drain piping, the total square footage conveying through each section must be determined.

 For example: How many square feet of drainage is flowing through Section A? **500 sq. ft.**

Using *Table 1106.3*, find the 3" rainfall rate column and the ⅛" slope. Locate the numerical values under the 3" column and find the one that will accommodate at least 500 sq. ft. of drainage. Once identified, follow that row to the left side of the table to determine the correct size of Section A horizontal piping.

Answer: 3"

Section B:

1. Conveying 500 sq. ft. of drainage from Leader 1 and 500 sq. ft. from Leader 2. Total drainage is 1,000 sq. ft.

Answer: 3"

Section C:

1. Conveying 500 sq. ft. of drainage from Leader 1, 500 sq. ft. from Leader 2, and 500 sq. ft. from Leader 3. Total drainage is 1,500 sq. ft.

Answer: 4"

Study Tip

Square footage of roof must be determined before sizing horizontal storm drains.

STORM DRAIN DIAGRAM

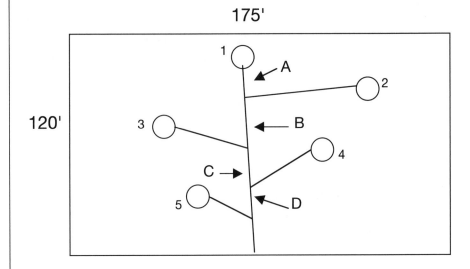

5" Rainfall rate per hour
$\frac{1}{4}$" slope

Leader Size: _____ inches

Horizontal A: _____ inches

Horizontal B: _____ inches

Horizontal C: _____ inches

Horizontal D: _____ inches

PLUMBING ISOMETRICS

The following exercise, based on Chapters 7 and 9 of the *International Plumbing Code*, will demonstrate how to properly size building drains, horizontal drains, and vents

Figure 3

	Water closet greater than 1 gpf
	Floor drain
	Double lavatory
	Urinal greater than 1.0 gpf
	Drinking fountain
	Laundry tray 2 compartments
	Washing machine 2" trap
------	Vent pipe

20 shower heads in shower room
Building drain = 1/4" slope

* This drawing may contain code errors

Drainage Fixture Units (Figure 3) Use Table 709.1 for dfu Value

	# Fixtures		dfu Value		Total
Water closets	_____	×	_____	=	_____
Urinals	_____	×	_____	=	_____
Lavatories	_____	×	_____	=	_____
Floor drains	_____	×	_____	=	_____
Shower heads	_____	×	_____	=	_____
Water fountains	_____	×	_____	=	_____
Washing machines	_____	×	_____	=	_____
Laundry tray	_____	×	_____	=	_____
			Total	=	_____

Horizontal Drain Sizing (Figure 3)

Tip: When calculating a drainage fixture unit total for a horizontal drain, make a list of each type of fixture on the branch. This will help ensure all fixture types are accounted for. From this point, an accurate tally of drainage fixture units can be determined to correctly size the horizontal drain.

	Fixture Type	# of		dfu Value	Total
Horizontal Branch A:	Water Closet	3	×	6	18
	Floor Drains	2	×	2	4
	Urinals	2	×	4	8
					= 30

Horizontal branch A has a total of 30 dfus. Using **Table 710.1(2)** an accurate drain size can be determined.

What is the minimum drain size for horizontal branch A? **Ans:** *4"*

Use Figure 3 on page 93 to successfully complete the rest of the worksheet.

	Fixture Type	# of	dfu Value	Total
Horizontal Branch B:	_____	____ ×	_____	= ____
	_____	____ ×	_____	= ____
	_____	____ ×	_____	= ____
			Total	= ____

	Fixture Type	# of	dfu Value	Total
Horizontal Branch C:	_____	____ ×	_____	= ____
	_____	____ ×	_____	= ____
	_____	____ ×	_____	= ____
	_____	____ ×	_____	= ____
			Total	= ____

	Fixture Type	# of	dfu Value	Total
Horizontal Branch D:	_____	____ ×	_____	= ____
	_____	____ ×	_____	= ____
			Total	= ____
Horizontal Branch E:	_____	____ ×	_____	= ____
	_____	____ ×	_____	= ____
			Total	= ____
Horizontal Branch F:	_____	____ ×	_____	= ____
	_____	____ ×	_____	= ____
			Total	= ____

Each branch dfu total can now be used to size the horizontal drains.

1. What is the minimum drain size for horizontal branch B? _____
2. What is the minimum drain size for horizontal branch C? _____
3. What is the minimum drain size for horizontal branch D? _____
4. What is the minimum drain size for horizontal branch E? _____
5. What is the minimum drain size for horizontal branch F? _____

Sizing Building Drain (Figure 3)

When sizing a specific section of a building drain, one must first determine the total number of drainage fixture units flowing into that particular zone of pipe. Using Table 710.1(1), choose the correct column according to the given slope and size that specified section of drain pipe.

Building Drain A–B: This zone is conveying all fixture units discharged from horizontal branch A: (30 dfus). The slope, given in figure 3, is ¼". Column 4 of Table 710.1(1) determines the maximum number of dfus allowed to connect to a certain diameter of pipe having a ¼" slope.

	Total dfu Value		**Building Drain Size**
Building Drain A–B:	30	=	3"
Building Drain B–C:	_____	=	_____
Building Drain C–D:	_____	=	_____
Building Drain D–E:	_____	=	_____
Building Drain E–F:	_____	=	_____

Vent Sizing (Figure 3)

IPC 2009, Table 916.1: This is the most common table used for sizing vents. The drain size and the number of fixture units being vented must be known in order to properly size the vent.

If you notice, in the first column, there are multiple listings of the same diameter of pipe. To ensure the correct diameter is chosen, the total fixture units being vented must be selected from column 2. The next step is to determine the developed length of vent pipe, which is usually given in the figure illustrated or the question. Everything needed to determine the vent size is in Table 916.1. Make sure you pay close attention to what each column represents.

	Drain Size	**# FU Vented**	**Developed Length**		**Vent Size**
Horizontal Branch A:	3"	30	30'	=	2"
Horizontal Branch B:	_____	_____	_____	=	_____
Horizontal Branch C:	_____	_____	_____	=	_____
Horizontal Branch D:	_____	_____	_____	=	_____
Horizontal Branch E:	_____	_____	_____	=	_____
Horizontal Branch F:	_____	_____	_____	=	_____

ISOMETRIC PRACTICE EXAM

Note: Drawings not to scale

Diagram A

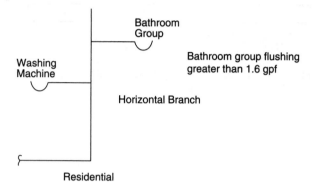

Refer to drawing above.

1. What is the total drainage fixture units?

 a. 4 dfu

 b. 6 dfu

 c. 10 dfu

 d. 8 dfu

2. What is the minimum size horizontal branch required?

 a. 1½"

 b. 3"

 c. 4"

 d. 2½"

Notes

Diagram B

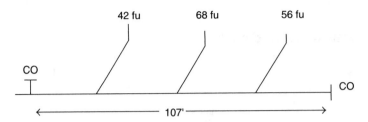

Refer to drawing above.

1. What size building drain is required if slope is ¼"?

 a. 4"

 b. 3"

 c. 2"

 d. 6"

2. What is wrong in this drawing as per most codes?

 a. 107' is more than maximum distance between cleanouts.

 b. There are too many branch lines.

 c. Nothing is wrong

 d. Drainage fixture units are not the same in each branch.

3. What size building drain is required if slope is ⅛"?

 a. 2"

 b. 3"

 c. 4"

 d. 6"

Diagram C

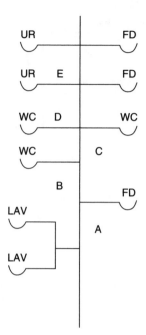

Urinals flushing greater than 1 gpf

Refer to drawing above.

1. What is the total drainage fixture units if WCs are tank type?

 a. 26 dfu

 b. 10 dfu

 c. 18 dfu

 d. 28 dfu

2. At what section of the horizontal branch would drain size increase from 2½" to 3"?

 a. C

 b. A

 c. F

 d. D

Notes

Diagram D

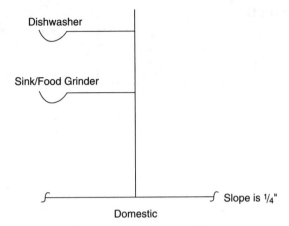

Refer to drawing above.

1. What would be the total drain fixture units (dfu) for the horizon-
 tal drain?

 a. 5 dfu

 b. 2 dfu

 c. 6 dfu

 d. 4 dfu

2. What size building drain is required in the above drawing?

 a. 1½"

 b. 3"

 c. 2"

 d. 4"

Diagram E

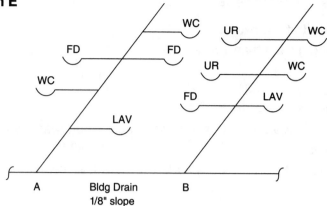

A Bldg Drain B
1/8" slope

Notes

Refer to drawing above.

1. What is the total drainage fixture units (WCs are public flushing greater than 1.6 gpf and urinals are flushing greater than 1 gpf)?

 a. 17 dfu

 b. 40 dfu

 c. 23 dfu

 d. 36 dfu

2. What is the minimum size drain required for horizontal branch A?

 a. 3"

 b. 4"

 c. 6"

 d. 2"

3. What is the minimum size drain required for horizontal branch B?

 a. 6"

 b. 3"

 c. 4"

 d. 2"

4. What is the minimum size building drain required?

 a. 3"

 b. 4"

 c. 6"

 d. 8"

Note: Use diagram F and F1 to answer Fixture Unit Practice Test.

- Tables are available from the IPC at the end of the exam.
- Assume all water closets are 'public' flushing greater than 1.6 gal.
- Urinals flushing greater than 1 gpf.
- The dishwasher and food grinder are installed on a 2" trap.
- The A/C condensate line at C12 is rated at 1 fixture unit.
- The five waste outlets in each shower room serve two shower heads each. Count each waste outlet in each shower room as 2 domestic showers.
- Unless otherwise stated, all drain lines are installed on a slope of ¼" per foot, all vents are less than 40 ft. long.
- Drainage plumbing under slab.

Diagram F

Diagram F1

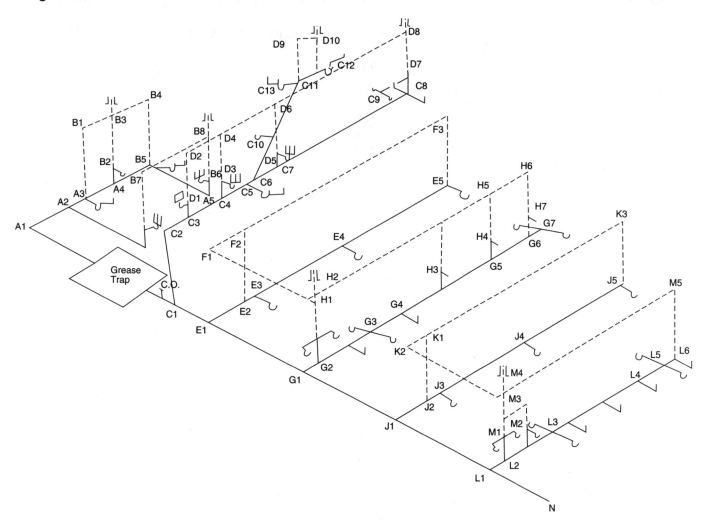

Practice Test for Fixture Units

1. What is the total drainage fixture unit (dfu) value at A1?

 a. 8 dfu

 b. 11 dfu

 c. 12 dfu

 d. 14 dfu

2. What is the drainage pipe size required for branch A1?

 a. 1½"

 b. 3"

 c. 4"

 d. 2½"

3. What is the total drainage fixture unit (dfu) value for C1?

 a. 28 dfu

 b. 20 dfu

 c. 22 dfu

 d. 32 dfu

4. What size drainage pipe is required between C9 and D7?

 a. 1¼"

 b. 2½"

 c. 2"

 d. 1½"

5. What type of fitting at C9? Fitting supports 2 lavs.

 a. Wye

 b. Sanitary tee

 c. Sanitary cross

 d. Double 90° elbow

6. How many dfu for branch G1?

 a. 25 dfu

 b. 38 dfu

 c. 42 dfu

 d. 32 dfu

7. What size cleanout is required by the grease trap?

 a. 2"

 b. 3"

 c. 4"

 d. 2½"

8. What size would be the vent pipe at B2?

 a. 1½"

 b. 1¼"

 c. 2"

 d. 2½"

Notes

9. What is the dfu count at J1?

 a. 3 dfu

 b. 6 dfu

 c. 15 dfu

 d. 12 dfu

10. How many dfu for branch L1?

 a. 41 dfu

 b. 39 dfu

 c. 34 dfu

 d. 36 dfu

11. What type of fitting is required at L4?

 a. 4" wye

 b. 3" sanitary tee

 c. 3" combo y and ⅛ bend

 d. 3" elbow

12. What is dfu count for vent through roof at B3?

 a. 6 dfu

 b. 4 dfu

 c. 8 dfu

 d. 11 dfu

13. What type of fitting is between trap arm M1 and M2? Fitting supports 2 lavs.

 a. 1½" cross

 b. 2" cross

 c. 1½" double sanitary tee

 d. 2" × 1½" double sanitary tee

14. What is the size of the building drain?

 a. 3"

 b. 6"

 c. 4"

 d. 8"

15. What would be the size of the vent through roof at M4 if developed length is 35 feet?

 a. 2"

 b. 2½"

 c. 3"

 d. 4"

16. What is trap size at E5 if each shower drain has 2 shower heads?

 a. 1½"

 b. 3"

 c. 2½"

 d. 2"

17. What is the size of vent through roof at H2 if developed length is 32 feet?

 a. 2"

 b. 2½"

 c. 1½"

 d. 3"

TABLE 709.1
DRAINAGE FIXTURE UNITS FOR FIXTURES AND GROUPS

Fixture type	Drainage fixture unit value as load factors	Minimum size of trap (inches)
Automatic clothes washers, commercial[a]	3	2
Automatic clothes washers, residential	2	2
Bathroom group consisting of water closet, lavatory, bidet, and bathtub or shower	6	—
Bathtub[b] (with or without overhead shower or whirlpool attachments)	2	1½
Bidet	2	1¼
Combination sink and tray	2	1½
Dental lavatory	1	1¼
Dental unit or cuspidor	1	1¼
Dishwashing machine,[c] domestic	2	1½
Drinking fountain	½	1¼
Emergency floor drain	0	2
Floor drains	2	2
Kitchen sink, domestic	2	1½
Kitchen sink, domestic with food waste grinder and/or dishwasher	2	1½
Laundry tray (1 or 2 compartments)	2	1½
Lavatory	1	1¼
Shower compartment, domestic	2	2
Sink	2	1½
Urinal	4	—[d]
Urinal, 1 gallon per flush or less	2[e]	—[d]
Wash sink (circular or multiple) each set of faucets	2	1½
Water closet, flushometer tank, public or private	4[e]	—[d]
Water closet, private installation	4	—[d]
Water closet, public installation	6	—[d]

For SI: 1 inch = 25.4 mm, 1 gallon = 3.785 L.
[a] For traps larger than 3 inches, use Table 709.2.
[b] A showerhead over a bathtub or whirlpool bathtub attachments does not increase the drainage fixture unit value.
[c] See Sections 709.2 through 709.4 for methods of computing unit value of fixtures not listed in Table 709.1 or for rating of devices with intermittent flows.
[d] Trap size shall be consistent with the fixture outlet size.
[e] For the purpose of computing loads on building drains and sewers, water closets or urinals shall not be rated at a lower drainage fixture unit unless the lower values are confirmed by testing.

TABLE 709.2
DRAINAGE FIXTURE UNITS FOR FIXTURE DRAINS OR TRAPS

Fixture drain or trap size (inches)	Drainage fixture unit value
1¼	1
1½	2
2	3
2½	4
3	5
4	6

For SI: 1 inch = 25.4 mm.

TABLE 710.1(1)
BUILDING DRAINS AND SEWERS

Diameter of pipe (inches)	Maximum number of drainage fixture units connected to any portion of the building drain or the building sewer, including branches of the building drain[a]			
	Slope per foot			
	¹⁄₁₆ inch	⅛ inch	¼ inch	½ inch
1¼	—	—	1	1
1½	—	—	3	3
2	—	—	21	26
2½	—	—	24	31
3	—	36	42	50
4	—	180	216	250
5	—	390	480	575
6	—	700	840	1,000
8	1,400	1,600	1,920	2,300
10	2,500	2,900	3,500	4,200
12	2,900	4,600	5,600	6,700
15	7,000	8,300	10,000	12,000

For SI: 1 inch = 25.4 mm, 1 inch per foot = 0.0833 mm/m.
[a]The minimum size of any building drain serving a water closet shall be 3 inches.

	Maximum number of drainage fixture units (dfu)			
TABLE 710.1(2) — **HORIZONTAL FIXTURE BRANCHES AND STACKS**[a]				
		Stacks[b]		
Diameter of pipe (inches)	**Total for a horizontal branch**	**Total discharge into one branch interval**	**Total for stack of three branch intervals or less**	**Total for stack greater than three branch intervals**
1½	3	2	4	8
2	6	6	10	24
2½	12	9	20	42
3	20	20	48	72
4	160	90	240	500
5	360	200	540	1,100
6	620	350	960	1,900
8	1,400	600	2,200	3,600
10	2,500	1,000	3,800	5,600
12	3,900	1,500	6,000	8,400
15	7,000	—[c]	—[c]	—[c]

For SI: 1 inch = 25.4 mm.

[a]Does not include branches of the building drain. Refer to Table 710.1(1).

[b]Stacks shall be sized based on the total accumulated connected load at each story or branch interval. As the total accumulated connected load decreases, stacks are permitted to be reduced in size. Stack diameters shall not be reduced to less than one-half of the diameter of the largest stack size required.

[c]Sizing load based on design criteria.

TABLE 916.1
SIZE AND DEVELOPED LENGTH OF STACK VENTS AND VENT STACKS

Diameter of soil or waste stack (inches)	Total fixture units being vented (dfu)	Maximum developed length of vent (feet)[a] Diameter of vent (inches)										
		1¼	1½	2	2½	3	4	5	6	8	10	12
1¼	2	30										
1½	8	50	150									
1½	10	30	100									
2	12	30	75	200								
2	20	26	50	150								
2½	42		30	100	300							
3	10		42	150	360	1,040						
3	21		32	110	270	810						
3	53		27	94	230	680						
3	102		25	86	210	620						
4	43			35	85	250	980					
4	140			27	65	200	750					
4	320			23	55	170	640					
4	540			21	50	150	580					
5	190				28	82	320	990				
5	490				21	63	250	760				
5	940				18	53	210	670				
5	1,400				16	49	190	590				
6	500					33	130	400	1,000			
6	1,100					26	100	310	780			
6	2,000					22	84	260	660			
6	2,900					20	77	240	600			
8	1,800						31	95	240	940		
8	3,400						24	73	190	720		
8	5,600						20	62	160	610		
8	7,600						18	56	140	560		
10	4,000							31	78	310	960	
10	7,200							24	60	240	740	
10	11,000							20	51	200	630	
10	15,000							18	46	180	570	
12	7,300								31	120	380	940
12	13,000								24	94	300	720
12	20,000								20	79	250	610
12	26,000								18	72	230	500
15	15,000									40	130	310
15	25,000									31	96	240
15	38,000									26	81	200
15	50,000									24	74	180

For SI: 1 inch = 25.4 mm, 1 foot = 304.8 mm.
[a]The developed length shall be measured from the vent connection to the open air.

PART FIVE

International Plumbing Code
Final Review Exam

Notes

INTERNATIONAL PLUMBING CODE
FINAL REVIEW EXAM

1. A laundry tray has an effective opening of 1½". What is the minimum size air gap required if away from the wall?

 a. 1.5"

 b. 2.5"

 c. 3"

 d. None of the above

2. Water hammer arresters shall conform to which of the following standards?

 a. ASSE 1010

 b. ASSE 1066

 c. ASSE 5020

 d. ASSE 5047

3. If a pump is rated 20 psi, what pressure is required to perform a forced sewer test?

 a. 5 psi

 b. 10 psi

 c. 20 psi

 d. 25 psi

4. Which of the following requires a cleanout to be installed in a building drain?

 a. 30° change in direction

 b. 35° change in direction

 c. 45° change in direction

 d. 90° change in direction

5. A water heater in an attic must have a clear access opening of _____.

 a. 20" × 30"

 b. 22" × 30"

 c. 24" × 30"

 d. 30" × 30"

6. What is the minimum size water-closet compartment allowed by code?

 a. 15"

 b. 21"

 c. 30"

 d. 1,800 sq. in.

7. Steel pipe shall be horizontally supported every _____ feet.

 a. 5

 b. 10

 c. 12

 d. 15

8. Which of the following is required to be a minimum distance of 25 feet away from an individual water supply?

 a. Barnyard

 b. Septic tank

 c. Sewer

 d. Pasture

Notes

Study Tip

Water hammer can be caused by sudden stop of water flow or vibration from the flow of water through pipe and fittings.

9. What is the minimum trap size for an emergency floor drain?

 a. 1¼"

 b. 1½"

 c. 2"

 d. None of the above

10. Standpipes shall be individually trapped and extend a minimum _____ inches and a maximum _____ inches.

 a. 18 / 42

 b. 18 / 36

 c. 42 / 18

 d. 42 / 36

11. A theater has a maximum occupant load of 500 people. What is the minimum number of water closets required for the female patrons?

 a. 2

 b. 3

 c. 4

 d. 8

12. A removable strainer from a shower waste outlet must be at least _____ inches in diameter.

 a. 2

 b. 3

 c. 4

 d. 5

13. A minimum ½ inch fixture supply pipe is required for all of the following except?

 a. Bathtub

 b. Dishwasher

 c. One-piece water closet

 d. Drinking fountain

14. What size horizontal branch is required to serve 620 drainage fixture units?

 a. 5"

 b. 6"

 c. 8"

 d. 10"

15. What is the minimum diameter of service pipe required to be installed in a residential service?

 a. ½"

 b. ¾"

 c. 1"

 d. 1½"

16. The minimum opening of a scupper to prevent the depth of ponding water from exceeding the roof's capability is _____ inches.

 a. 2

 b. 3

 c. 4

 d. 5

Study Tip

A pasture must be a minimum of 100 feet away from a private water supply.

Study Tip

An emergency floor drain has a fixture unit count of 0.

17. What is the maximum water pressure allowed within a building?

 a. 80 psi

 b. 100 psi

 c. 140 psi

 d. 160 psi

18. <u>Given:</u> Drain size = 4 inches

 Fixture units being vented = 260
 Developed length of vent = 60 feet
 What is the minimum size vent stack required?

 a. 1½"

 b. 2"

 c. 2½"

 d. 3"

19. A drainage and vent air test is performed until there is a uniform gauge pressure of 5 psi and must be maintained a minimum of _____.

 a. 5 minutes

 b. 10 minutes

 c. 15 minutes

 d. 20 minutes

20. When backfilling, the crown of the pipe must be covered by _____ of tamped earth.

 a. 6"

 b. 12"

 c. 18"

 d. 24"

21. What is the maximum occupant load allowed before separate facilities are required?

 a. 2

 b. 4

 c. 10

 d. 15

22. What is the minimum clearance in front of a water closet to any wall?

 a. 15"

 b. 21"

 c. 30"

 d. 60"

23. What is the fixture unit value for an emergency floor drain?

 a. 0

 b. 1

 c. 2

 d. 4

24. What is the minimum size wet vent permitted to be used?

 a. 1½"

 b. 2"

 c. 2½"

 d. 3"

25. Which of the following is detrimental to the sewer system?

 a. Ashes

 b. Cinders

 c. Grease

 d. All of the above

Study Tip

A scupper is a secondary emergency roof drain.

Notes

PART SIX
International Fuel Gas Code

The following section covers the types of questions that are likely to be asked on most plumbing exams that include the *International Fuel Gas Code* as a reference.

Study Tip

The required water closets, lavatories, and showers or bathtubs shall be distributed equally between the sexes based on the percentage of each sex anticipated in the occupant load.

INTERNATIONAL FUEL GAS CODE PRACTICE EXAM

1. What is the minimum burial depth required for an individual gas line connected to a grill?

 a. 8"

 b. 12"

 c. 18"

 d. 24"

2. If a 1-inch pipe needs to be field threaded, what is the approximate number of threads to be cut?

 a. 8

 b. 10

 c. 12

 d. 13

3. What is the approximate gas input for a built-in oven installed in a single-family dwelling?

 a. 65,000 Btu/h

 b. 40,000 Btu/h

 c. 25,000 Btu/h

 d. 3,000 Btu/h

4. Which of the following is the standard weight for steel and wrought-iron pipe?

 a. Schedule 40

 b. Schedule 80

 c. Schedule 120

 d. All of the above

5. Which of the following materials is used to make a flange gasket?

 a. Metal-jacketed asbestos

 b. Aluminum O-ring

 c. Asbestos

 d. All of the above

6. The maximum amount of phosphorous allowed in brazing alloy is _____.

 a. 0.02%

 b. 0.04%

 c. 0.05%

 d. 0.08%

7. Appliance fuel connectors shall have an overall length not to exceed _____ feet.

 a. 4

 b. 5

 c. 6

 d. 10

8. ¾-inch gas tubing is required to be supported every _____.

 a. 4'

 b. 6'

 c. 8'

 d. every floor level

Study Tip

The minimum burial depth of a gas line is 12".

9. Medium pressure regulators must comply with all of the following except _____.

 a. a tee fitting with one opening capped shall be installed between the MP regulator and the upstream shutoff valve

 b. an MP regulator shall maintain a reduced outlet pressure under lock-up conditions

 c. an MP regulator must be checked for leaks once every six months or according to the manufacturer's recommendation

 d. an MP regulator shall be suitable for the inlet and outlet gas pressures

10. What is the maximum depth a joist can be notched on the ends?

 a. 20%

 b. 25%

 c. 33%

 d. 40%

11. For compressed natural gas motor vehicle fuel-dispensing facilities, an emergency shutdown device shall be located not less than _____ feet from dispensers and shall also be provided in the compressor area.

 a. 25

 b. 50

 c. 75

 d. 100

12. Given: Natural Gas, Inlet Pressure of 0.5 psi, Pressure drop is 0.5" W.C. (Schedule 40 Metallic Pipe)

 A 50-foot pipe is connected to a barbeque. What size gas pipe is required?

 a. ½"

 b. ¾"

 c. 1"

 d. 1¼"

13. What is the maximum operating pressure for a piping system located in a building?

 a. 5 psi

 b. 10 psi

 c. 15 psi

 d. 20 psi

14. In non-load-bearing partitions, any wood stud is permitted to be notched; however, the maximum depth allowed shall not exceed _____.

 a. 20%

 b. 25%

 c. 40%

 d. 60%

15. The quantity of heat required to raise 1 pound of water 1°F is known as a _____.

 a. demand

 b. watt

 c. Btu

 d. None of the above

Study Tip

Silver is used for brazing all ferrous and nonferrous metals.

Study Tip

A load bearing wall may be notched a maximum of 25% the depth of the stud.

GAS PIPE SIZING QUESTIONNAIRE

Step 1: Find the longest run of pipe from the point of delivery (meter) to the most remote outlet.

Once that length is determined, look at the left-hand column of Table 402.4(2) and find that run of pipe. If the length falls in between two of the numbers, choose the higher number. (Number in the column must be equal to or greater than your longest run of pipe.)

When the correct length is found in the table, every section of pipe will be sized according to that row. All other lengths of pipe remaining are no longer important.

Step 2: Convert Btu/h into cubic feet of gas Btu ÷ 1,000

Example: Convert 35,000 Btu to cubic feet of gas
 Ans: 35,000 ÷ 1,000 = 35 cubic feet of gas

Step 3: You are now ready to size gas pipe. Find the longest run of gas pipe and size every section according to that length. The numbers in the middle of the table represent cubic feet of gas. Study the table carefully and you will realize that everything you need to know to use it correctly is right before your eyes.

Gas Pipe Sizing—IFG Table 402.4(2)
(Figure 4)

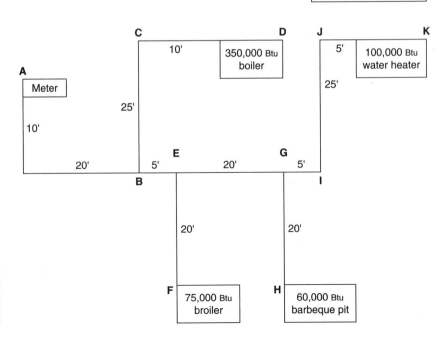

Questions 1–10 refer to Figure 4.

1. What is the minimum size gas pipe allowed to serve section A–B?

 a. ¾"

 b. 1"

 c. 1¼"

 d. 1½"

2. What is the minimum size gas pipe allowed to serve section B–C?

 a. ¾"

 b. 1"

 c. 1¼"

 d. 1½"

3. What is the minimum size gas pipe allowed to serve section C–D?

 a. ¾"

 b. 1"

 c. 1¼"

 d. 1½"

4. What is the minimum size gas pipe allowed to serve section B–E?

 a. ½"

 b. ¾"

 c. 1"

 d. 1¼"

5. What is the minimum size gas pipe allowed to serve section E–F?

 a. ⅜"

 b. ½"

 c. ¾"

 d. 1"

Study Tip

Btu = British Thermal Unit.

Study Tip

The longest length method is the most common method used for sizing gas piping.

6. What is the minimum size gas pipe allowed to serve section E–G?

 a. ⅜"

 b. ½"

 c. ¾"

 d. 1"

7. What is the minimum size gas pipe allowed to serve section G–H?

 a. ⅜"

 b. ½"

 c. ¾"

 d. 1"

8. What is the minimum size gas pipe allowed to serve section G–I?

 a. ½"

 b. ¾"

 c. 1"

 d. 1¼"

9. What is the minimum size gas pipe allowed to serve section I–J?

 a. ½"

 b. ¾"

 c. 1"

 d. 1¼"

10. What is the minimum size gas pipe allowed to serve section J–K?

 a. ½"

 b. ¾"

 c. 1"

 d. 1¼"

PART SEVEN
General Knowledge

A portion of your exam will be based on general knowledge. These questions will be based on plumbing theory and the answers will be found in an approved textbook rather than the codebook. In identifying questions that are based on general knowledge, remember the codebook tells you what you must do and does not indicate how something is done. If the question includes the words *minimum*, *maximum*, *must be*, or *shall be,* it is probably a code question. If the question is based on how and why something is done, the answer will be found in the approved reference book that discusses plumbing technology and theory.

GENERAL KNOWLEDGE PRACTICE EXAM #1

1. What is/are the required fixture/s for a family dwelling?

 a. Water closet

 b. Lavatory

 c. Tub/shower and sink

 d. All of the above

2. Plumbing systems in buildings include each of the following EXCEPT:

 a. Potable water

 b. Sanitary drainage

 c. Fuel gas piping

 d. Storm water

3. When planning a plumbing system, which of the following should be considered with regard to the piping?

 a. Available materials

 b. Local plumbing codes

 c. Cost of material, ease of installation, weight, and durability of material

 d. All of the above

4. Name the four types of copper tube.

 a. Types schedule 80; schedule 40; schedule 30; schedule 10

 b. Types K; L; M; DWV

 c. Types heavy duty; service weight; with coupling; without coupling

 d. Types underground; above ground; in-wall; thin wall

5. What is galvanic action?

 a. Galvanic action is an electrolytic chemical reaction between dissimilar metals such as copper and iron

 b. Volcano action from a mountain or hill

 c. When sewer water enters a potable water system

 d. None of the above

6. What are the most common scales used in building drawings?

 a. $\frac{1}{4}$" = 1' and $\frac{1}{2}$" = 1'

 b. $\frac{1}{8}$" = 1" and $\frac{1}{4}$" = 1'

 c. $\frac{1}{8}$" = 1' and $\frac{1}{4}$" = 1'

 d. $\frac{1}{8}$" = 1' and $\frac{3}{8}$" = 1'

 e. All of the above

 f. None of the above

7. What does this mean: SCALE $\frac{1}{4}$" = 1'0"

 a. Each $\frac{1}{4}$" of measurement equals 1 foot

 b. Each $\frac{1}{4}$" of measurement equals 1 inch

 c. Each 1 inch equals 4 feet

 d. Both a and c

8. Wall-hung water closet bowls shall be _____.

 a. bolted securely to horizontal blocking

 b. connected by the supplied bolts to the wall

 c. rigidly supported by concealed metal supports

 d. None of the above

Study Tip

A bathroom receptacle must be GFI unless a lavatory is not present.

9. Plumbing codes are area laws established to control installation, maintenance, removal, and safety of plumbing conditions for the public and the plumbers or other craftsmen in the _____.

 a. city

 b. county

 c. state

 d. All of the above

10. Who issues licenses?

 a. Plumbing Inspector

 b. State and/or local authorities

 c. I.R.S.

 d. City Clerk

11. Each of the following is a requirement for plumbing fixtures EXCEPT:

 a. They shall be smooth surfaced

 b. They shall be made of non-absorbent material

 c. They shall be made of unbreakable material

 d. They shall be free of defects

12. What type of plugs do underground wyes and test tees use?

 a. Cast iron plugs

 b. Brass plugs

 c. Inflatable rubber test plugs

 d. Screw plugs

13. What does a testing gauge assembly do?

 a. It shows the pressure value to guarantee that your system is maintaining the rating required by that particular test

 b. It shows the required feet needed to give sufficient pressure

 c. It shows the slope needed to properly drain the system

 d. It shows the system being checked will pass the plumbing codes

14. What do the kitchen sink, garbage disposal, and dishwasher share?

 a. Single trap and drainpipe

 b. Water line

 c. 1¼" drain

 d. All of the above

15. What is a license?

 a. Permission for the plumber to begin working alone

 b. Displays their knowledge, experience, and skill levels

 c. Both a and b

 d. You do not need a license to do plumbing

16. All water piping conveying nonpotable water shall be adequately and durably identified by a distinct _____ colored paint.

 a. red

 b. blue

 c. yellow

 d. None of the above

Study Tip

Galvanic action is caused from a small electrical current being generated between two metals.

Study Tip

DWV = Drainage, Waste, and Vent Piping.

17. A _____ is a document that grants permission for an individual or company to do plumbing work at a predetermined site or building.

 a. license

 b. contract

 c. permit

 d. certificate of occupancy

18. What are the blueprint plans that provide information about the plumbing, heating, and air conditioning systems of a building called?

 a. Plumbing

 b. Sepia

 c. Mechanical

 d. Architectural

19. What are some ways contamination can occur?

 a. Gravity backflow and cross-connections

 b. backsiphonage and backpressure

 c. Both a and b

 d. None of the above

20. On a gas water heater, what is a thermocouple?

 a. A small electric generator

 b. Fuel supply for pilot

 c. Water sensor

 d. Regulates temperature

21. A plumbing fitting which makes a 60° bend would also be known as a(n):

 a. ⅛ bend

 b. ⅙ bend

 c. ⅕ bend

 d. ¼ bend

22. What do you do if water hammering occurs in your water heater?

 a. Install a water hammer arrester

 b. Install an expansion tank

 c. Install an air chamber

 d. Either a, b, or c

23. The use of primer and cement to make a joint in PVC pipe is known as:

 a. Solvent welding

 b. Insert fittings with hose and clamp

 c. Air welding

 d. Insert fittings with clamp rings

24. What is the difference between a sink and a lavatory?

 a. Sinks are used as washing fixtures and a lavatory is a functional unit.

 b. Sinks are functional units and a lavatory is used as a washing fixture

 c. Sinks have three compartments and a lavatory has one compartment.

 d. None of the above

Study Tip

A domestic garbage disposal must have a minimum trap size of 1½".

Study Tip

The set point temperature of a water tank determines whether the energy supply needs to be turned on or off.

25. What does ECO stand for when related to a water heater?

 a. Emergency control orifice

 b. Energy cutoff

 c. Both a and b

 d. None of the above

GENERAL KNOWLEDGE PRACTICE EXAM #2

1. What is an inspection?

 a. Inspections guarantee that the work fulfills code requirement and that the work is professional. An inspector will issue an "accepted" or "rejected" certificate

 b. Inspections give individuals permission to perform plumbing within the jurisdiction of authority granting the inspection

 c. Inspections make sure plumbers are doing quality work and that it is done according to code

 d. Both a and c

2. Which plumbing systems require testing?

 a. Potable water supply, distribution systems, and water services

 b. Storm drainage systems outside buildings

 c. Sanitary and venting systems, storm drainage systems inside buildings

 d. All of the above

 e. Both a and c

3. How are drainage and venting in buildings larger than 5 stories high tested?

 a. One test for all stories is O.K.

 b. Test each 5-story section separately with a 10' head of water.

 c. Test each story with a 10' head of water.

 d. Test all stories with 10 psi.

4. What prevents a chemical sink from being damaged?

 a. Continually running water

 b. Manufactured out of chemical-resistant materials

 c. P-trap seal

 d. Both b and c

 e. Both a and b

Study Tip

Solvent cement joints shall be permitted above or below ground.

Study Tip

Insulating hot-water pipes from the water heater to each fixture will reduce energy consumption and more quickly supply hot water to, a faucet.

5. What do you have to do to a new garbage disposer before a dishwasher can be connected to it?

 a. Install a studor vent

 b. Install an air gap

 c. Knock out the drain plug located in the disposer drain outlet

 d. Nothing

6. What is the purpose of an anode rod on a water heater?

 a. The magnesium rod extends the tank life

 b. Makes the water hotter

 c. Controls the temperature

 d. Relieves pressure for safety

7. What is the purpose of a dip tube on a water heater?

 a. Transfers water from inlet to bottom of tank

 b. Filters out minerals from water

 c. Helps prevent backsiphonage by having a small hole at the top of the dip tube

 d. Both a and c

8. The desired air test pressure required for testing water supply lines:

 a. 1½ times the working pressure or 175 psi, whichever is the greater

 b. 2 times the working pressure or 175 psi, whichever is the larger

 c. Working pressure or 100 psi, whichever is the greater

 d. 1½ times the working pressure or 150 psi, whichever is the larger

9. The type of heating element used in electric water heaters:

 a. Low-voltage elements

 b. 24-volt heating elements

 c. Resistance heating elements

 d. All of the above

10. A 160-foot sewer pipe installed with 1% grade will fall ____ inches.

 a. 60

 b. 19

 c. 28¾

 d. 30⅝

11. What should you do before connecting electrical wires or turning on the electrical breaker to an electric water heater?

 a. Check to see if connections are watertight

 b. Make sure all air is out of the tank by opening a hot-water faucet and bleeding the air

 c. Make sure the water heater is level

 d. All of the above

12. What is a person called who has completed apprenticeship training in a particular field?

 a. Master

 b. Contractor

 c. Journeyman

 d. Worker

13. The primary purpose of a vent line in a water closet is _____.

 a. to remove odor from the building

 b. to prevent loss from trap seal

 c. to mix air with the sewage to help with the treatment

 d. to transport black water

Study Tip

The maximum height above the floor for an ADA approved sink is 34".

Study Tip

An air gap is the vertical distance from the top of the flood rim to the faucet. Prevents backflow.

14. Why is a relief valve so important to a water heater?

 a. It plugs a hole in the side/top of water heater

 b. It serves as a drain for the water heater

 c. It prevents super-heated water from exploding

 d. Both a and b

15. What are the two primary water sources for public potable drinking water?

 a. Potable water and contaminated water

 b. Surface water and ground water

 c. Rain water and well water

 d. All of the above

16. What is a cross-connection?

 a. When two separate piping systems one potable and one undesirable flow into each other

 b. When two dissimilar metals touch each other

 c. When two pipes cross

 d. None of the above

17. What is the hydrologic cycle?

 a. The continuous movement of water from the ocean, through the atmosphere, onto the land and back to the ocean

 b. The cycle completed when water is treated at the water works plant

 c. The chemical action when the rain enters the clouds

 d. Both a and b

18. Surface water enters the _____ through precipitation and infiltration.

 a. streams

 b. shallow wells

 c. lakes and reservoirs

 d. All of the above

19. What is ground water?

 a. Water that runs off the ground

 b. Water that is brought to surface with wells and pumps

 c. Water located deep within the earth

 d. Both b and c

20. What is the chemical agent that softens and cleans the surface of plastic pipe and allows the cement to penetrate more effectively?

 a. Cement

 b. Primer

 c. Glue

 d. Pipe compound

21. Pipes passing through the foundation walls should be _____.

 a. caulked

 b. damp proofed

 c. sleeved with cast iron pipe

 d. None of the above

Study Tip

Apprentice plumbers is a stage lasting generally 5 years.

Study Tip

Temperature/pressure relief valve opens to permit hot water to escape and cold water to enter the tank.

22. What is used in soldering and brazing to facilitate the removal of oxides or undesirable substances from copper pipe?

 a. Flux

 b. Sandpaper

 c. Emery cloth

 d. Acid

23. In some areas, faucets installed after 1990 are required to be:

 a. Thermostatic or pressure-balanced

 b. Cartridge

 c. Single handle

 d. All of the above

24. When calculating single-pipe offsets, what is the vertical distance between the centers of the two offset fittings called?

 a. Travel

 b. Offset

 c. Advance

 d. None of the above

25. What device can be used to add a small amount of water to a floor drain trap to ensure an adequate trap seal?

 a. Ball valve

 b. Trap primer

 c. Trap valve

 d. Check valve

GENERAL KNOWLEDGE PRACTICE EXAM #3

1. The grade of a drainage pipe is _____.

 a. the weight of the material

 b. the slope of the pipe

 c. the material sketch

 d. the abrasion resistance of the pipe

2. A temperature of at least _____ is necessary to braze joints and add the filler.

 a. 840°F

 b. 1,120°F

 c. 640°F

 d. 460°F

3. The inside diameter of a ¾ inches type L copper plumbing tube is:

 a. 0.811"

 b. 0.745"

 c. 0.785"

 d. 0.875"

4. What kind of pipe is joined together using a stainless steel clamp assembly surrounding a neoprene sleeve that seals the joint?

 a. PVC

 b. No-hub cast iron

 c. Hubbed cast iron

 d. None of the above

5. Which of the following is necessary for the purpose of directing water to the showerhead at tub and shower faucets?

 a. Cartridge

 b. Diverter

 c. Faucet washer

 d. Tub spout

Study Tip

Potable water means safe for human consumption.

Study Tip

A ball valve opens and closes with just one-quarter turn which provides very little resistance to flow.

6. What agency requires all employers to provide a safe environment for their employees?

 a. Environment Safety Association

 b. OSHA

 c. State Health & Safety Commission

 d. U.S. Board of Safety

7. A three-dimensional drawing of a plumbing system is referred to as:

 a. Schematic

 b. Floor plans

 c. Isometric

 d. Plan views

8. The most common individual vent is called:

 a. Straight

 b. Common

 c. Continuous

 d. Single

9. A vertical pipe installed to provide circulation of air to and from the drainage system is called:

 a. Vent stack

 b. Stack vent

 c. Main vent

 d. Main stack

10. What type of faucet uses a rubber washer pressing down on a brass seat to control the flow of water?

 a. Reverse compression

 b. Compression

 c. Balancing control

 d. Cartridge

11. What is the drain fixture unit rating for a kitchen sink?

 a. 2 wsfu

 b. 2 dfu

 c. 1 dfu

 d. 3 wsfu

12. The fitting that mounts under the drain plug on a bathtub waste and overflow is called:

 a. Drain fitting

 b. Waste boot

 c. Tub elbow

 d. Waste shoe

13. The Energy Policy Act of 1992 mandated that the maximum volume of water used to flush a water closet could not exceed how many gallons?

 a. 2.1 gal.

 b. 3.5 gal.

 c. 1.6 gal.

 d. 6.1 gal.

Study Tip

ABS stands for acrylonitrile butadiene styrene.

14. What is a valve called that has a flush device actuated by direct water pressure to supply a fixed quantity of water for flushing purposes?

 a. Flushometer

 b. Ballcock

 c. Vacuum breaker

 d. Flushvalve

15. What is the drainage pipe extending horizontally from a soil or waste stack or building drain, with or without vertical sections or branches?

 a. Main drain

 b. Horizontal branch drain

 c. Building sewer

 d. None of the above

16. Approximately how much does one cubic foot of water weigh?

 a. 7.5 lbs.

 b. 26½ lbs.

 c. 36 lbs.

 d. 62½ lbs.

17. What provides a liquid seal in drainage piping to prevent the escape of sewer gases into the living area?

 a. Trap

 b. Fixture unit

 c. Vent system

 d. None of the above

18. What is the most commonly used type of pipe and fittings today?

 a. Cast iron

 b. Copper

 c. PVC

 d. Steel

19. The distance between the crown weir and the top of the trap is:

 a. Crown length

 b. Weir distance

 c. Trap seal

 d. None of the above

20. What is caused when pressure in the water distribution pipe becomes higher than the water supply pressure?

 a. Backpressure

 b. Backsiphonage

 c. Cross-connection

 d. Atmospheric pressure

21. What is an air gap?

 a. Area between two pipes

 b. Atmospheric air between land and sky

 c. Distance between overflow tube and top of tank

 d. Unobstructed vertical space between the water supply and the flood rim of the receptacle

22. Expressed in inches, what is the distance the drainage pipe drops in its given length?

 a. Slope

 b. Fall

 c. Grade

 d. Both a and c

Study Tip

A trap seal is the vertical distance between the weir and the top of the dip of the trap.

23. What is the faucet installed on the outside of a building called?

 a. Check valve

 b. Sillcock

 c. Pressure vacuum breaker

 d. Water shutoff valve

24. What is the valve with hose threads installed to drain a water heater called?

 a. Boiler drain

 b. Hydrant

 c. Hose bibb

 d. Water shutoff

25. A valve that permits fluid flow in only one direction and closes automatically to prevent backflow is referred to as:

 a. Check valve

 b. Anti-siphon valve

 c. Ball valve

 d. All of the above

GENERAL KNOWLEDGE PRACTICE EXAM #4

1. What is a valve that converts high and/or fluctuating inlet water pressure to a lower or constant outlet pressure called?

 a. High-pressure valve

 b. Backpressure valve

 c. Pressure-reducing valve

 d. Both a and c

2. What type of traps do water closets have?

 a. P-trap

 b. Integral

 c. Bell trap

 d. Crown weir trap

3. In a flush tank assembly, how high is the water level in comparison to the overflow tube?

 a. The level should be ¾" below top of overflow tube

 b. The level should be at the top edge of overflow tube

 c. The level should be 1½" below top of overflow tube

 d. The level makes no difference.

4. In a flush tank assembly, what is the purpose of the ballcock assembly?

 a. Acts as an anti-siphon valve

 b. Fills tank with water

 c. Lets in the water

 d. All of the above

Study Tip

A cross-connection is a physical connection between two otherwise separate piping systems.

Study Tip

Check valves permit flow in only one direction.

5. What is the maximum flush volume for a urinal?

 a. 1.0 gpm

 b. 1.6 gpm

 c. ¾ gpm

 d. 3.5 gpm

6. PE pipe (polyethylene) is black and is typically sold:

 a. In 10' lengths

 b. By the foot

 c. In coils

 d. In 21' lengths

7. PE pipe (polyethylene) is used for:

 a. Jet wells and water supply systems

 b. Farm sprinkler systems

 c. Salt lines, chemical well lines, and gas gathering systems

 d. All of the above

8. How is PE (polyethylene) joined together?

 a. Plastic insert fittings and clamps

 b. No-hub clamps

 c. Cement and primer

 d. Compression couplings

9. Typically, manufacturers guarantee PVC pipe for:

 a. 10 years

 b. 50 years

 c. 75 years

 d. 100 years

10. One pint of plastic pipe cement will fuse approximately how many 1½-inch joints?

 a. 35

 b. 25

 c. 100

 d. 50

11. What color cement is used for CPVC pipe and fittings?

 a. Clear

 b. Orange

 c. Black

 d. Blue

12. What is the color difference in PVC and ABS pipe and fittings?

 a. PVC is black and ABS is white

 b. PVC is white and ABS is black

 c. PVC is plastic and ABS is cast iron

 d. PVC is schedule 40 and ABS is schedule 30

13. Can both PVC and ABS pipe and fittings be joined together with PVC cement?

 a. No, they are different types of plastic pipe

 b. Yes, PVC cement will work for any type of plastic

 c. Yes, but only for these two types of plastic

 d. Yes, as long as you use the correct primer

14. Can PVC pipe and fittings be used for the hot-water system?

 a. Yes, PVC can be used on hot or cold-water systems

 b. No, use CPVC or pipe that is approved for hot-water systems

 c. No, use only copper or galvanized steel pipe

 d. None of the above

Study Tip

Some ballcocks have a flexible tube that must be clipped to the flush valve overflow so that the water can enter the bowl of the toilet to fill the trap.

Study Tip

Most plastic water supply lines are measured using the face-to-face technique, with an allowance for the depth of the fitting socket.

15. What is the quietest drain pipe to install in a building?

 a. PVC plastic pipe

 b. Copper DWV pipe

 c. Steel pipe

 d. Cast iron pipe

16. The drawing scale of a drawing is ⅛" = 1'0". The length of the pipe measures 2¼" on the blueprint. How long is the actual length of the pipe?

 a. 16 ft.

 b. 18 ft.

 c. 18 in.

 d. 20 in.

17. On blueprints, what is the purpose of phantom lines?

 a. Used to display water lines

 b. Used to display buried pipes

 c. Used to display drainage and vent lines

 d. Used to display moveable parts

18. On a blueprint, what does a "key" tell you?

 a. Keys describe what the lines or specific abbreviations represent

 b. Keys tell where the fixtures are located

 c. Keys show where to install drainage pipes

 d. All of the above

19. The inside diameter of a 1½", schedule 40, nominal pipe size is:

 a. 1.90"

 b. 0.145"

 c. 1.610"

 d. 0.200"

20. A 3-inch short sleeve turn may be used in all but which of the following locations?

 a. In waste lines with a change in flow in horizontal to vertical

 b. In waste lines with a change in flow in vertical to horizontal

 c. In waste lines between floors

 d. In waste lines where the change in flow is from horizontal to horizontal

21. Ejector pumps may be connected to a _____.

 a. wye fitting in the building drain

 b. cleanout at the upstream end of the building sewer

 c. sanitary sewer and the building drain

 d. tee fitting and a branch of the building drain

22. Single sanitary tees, double sanitary tees, and quarter bends, may be used in vertical sections of drainage lines only where the direction of the flow is _____.

 a. from horizontal to vertical

 b. from vertical to horizontal

 c. less than 20° off horizontal

 d. less than 45° off horizontal

23. Drops for lavatories are made up to be ___ inches long.

 a. 24

 b. 30

 c. 36

 d. 42

Study Tip

Polyethylene (PE) is used most frequently to pipe natural gas.

Study Tip

Phantom lines on blueprints are used to display drainage and vent lines.

24. The blowout water closet is a commercial type water closet that depends on a large jet of water directed into the _____ of the trap.

 a. inlet

 b. outlet

 c. weir

 d. None of the above

25. When threading 1" steel pipe, the recommended number of threads per inch is:

 a. 11

 b. 11.25

 c. 11.5

 d. 11.75

GENERAL KNOWLEDGE PRACTICE EXAM #5

1. Which are typically used to fasten or secure pipe supports to structural components of wood frame construction?

 a. Plastic inserts

 b. Expansion plugs

 c. Nuts and bolts

 d. Nails and screws

2. Trap siphonage and backsiphonage pressure are a result of pressure differences between the drainage system and _____.

 a. ventilation duct

 b. extension trap

 c. biosphere

 d. atmosphere

3. Branch lines which enter a stack at or near each floor level in a building is known as a _____.

 a. vent stack

 b. waste stack

 c. branch interval

 d. branch soil pipe

4. Which of the following can only exist where there is another floor above the one in question?

 a. Wet vent

 b. Loop vent

 c. Circuit vent

 d. Yoke vent

Study Tip

Double sanitary tee patterns shall not receive the discharge of back-to-back water closets.

Study Tip

Threads on iron pipe fittings are tapered so they will form a watertight joint when tightened securely.

5. Which of the following are plumbing pipes made from?

 a. Plastic

 b. Copper

 c. Cast iron soil pipe

 d. All of the above

6. Water distribution pipes are generally represented by which color?

 a. Green

 b. Yellow

 c. Tan

 d. None on the above

7. A vent pipe that connects at the junction of two fixture drains and serves as a vent is known as a(n)_____.

 a. relief vent

 b. back vent

 c. common vent

 d. individual vent

8. A vent stack is connected to the waste stack at the lower end with a wye and _____.

 a. $\frac{1}{16}$ bend

 b. $\frac{1}{8}$ bend

 c. $\frac{1}{6}$ bend

 d. $\frac{1}{4}$ bend

9. In a multi-story building, the waste stack and vent stack must reconnect with a yoke vent every _____ branch intervals.

 a. 4

 b. 5

 c. 6

 d. 8

10. Strainer baskets attached to a building storm drain to prevent debris from entering storm water systems are typically made of _____.

 a. cast iron

 b. cast aluminum

 c. plastic

 d. All of the above

11. A trap primer is a device which adds a small amount of water to the floor drain trap to ensure an adequate trap seal. Which of the following is not a type of trap primer?

 a. Dispensing trap primer

 b. Automatic trap primer

 c. Water saver trap primer

 d. Vacuum breaker trap primer

12. In a multi-family dwelling, common fixtures are installed _____.

 a. back to back

 b. side to side

 c. front to back

 d. front to front

13. Which of the following is a prohibited trap if the inlet and outlet of the trap are in vertical alignment with each other?

 a. Gag

 b. Bag

 c. Slug

 d. Key

Study Tip

A vacuum is any pressure less than atmospheric pressure.

Study Tip

Stacks that are offset 45° or less from the vertical are sized as though they were straight vertical stacks.

14. What factor can be affected by a plumber when installing a plumbing system?

 a. Number of fixtures

 b. Size of piping

 c. Type of fittings

 d. Length of piping

15. Which of the following receives only liquid body waste and conveys the waste through a trap seal into a sanitary drainage system?

 a. Water closet

 b. Bidet

 c. Lavatory

 d. Urinal

16. How many bolts are required to anchor a siphon jet water closet bowl to the supporting chair carrier?

 a. 2

 b. 3

 c. 4

 d. 5

17. A floor set basin used for maintenance tasks and cleaning is known as a_____.

 a. floor sink

 b. utility sink

 c. maintenance basin

 d. mop basin

18. Which of the following physical properties affect water in supply lines?

 a. Density

 b. Boiling point

 c. Viscosity

 d. All of the above

Study Tip

Trap primer is a small amount of water added to the trap to ensure adequate trap seal.

19. A lawn sprinkler is connected to the potable water service supply pipe; there must be an approved backflow preventer on the discharge side of each valve. The backflow preventer must be at least _____.

 a. 6 inches above the highest sprinkler head

 b. 6 inches away from the highest sprinkler head

 c. 8 inches above the highest sprinkler head

 d. 8 inches away from the highest sprinkler head

20. Which of the following must include a yard hydrant because of their size, isolation, or type of occupancy?

 a. Mobile home park

 b. Marinas

 c. Lumber yards

 d. All of the above

21. Pedestal urinals must have a minimum of _____ inches center-to-center spacing when set in a battery installation.

 a. 15

 b. 18

 c. 30

 d. None of the above

22. Plastic piping from a septic tank to a drainfield located under paving shall be?

 a. ABS schedule 40

 b. PVC schedule 40

 c. PVC schedule 80

 d. CPVC piping only

23. On a loop system a vent shall be installed vertically downstream, and be within _____ feet of the first fixture branch.

 a. 5

 b. 7

 c. 8

 d. 10

24. Which of the following materials shall not be used for storm drainage within a building?

 a. Bituminized fiber pipe

 b. Galvanized steel pipe

 c. Plastic pipe

 d. Lead pipe

25. Which of the following hose-connected faucets are not required to have a backflow preventer installed?

 a. Three-compartment commercial sink

 b. Service sink

 c. Automatic clothes washing machine

 d. Outside hose faucet

GENERAL KNOWLEDGE PRACTICE EXAM #6

1. The process of joining two pieces of metal using heat and the application of another metal of different composition is called _____.

 a. brazing

 b. soldering

 c. compression

 d. face feeding

2. Capped pipe nipples that penetrate walls and floors to help locate supply and drain lines are known as _____.

 a. cleanouts

 b. location stubs

 c. stub pieces

 d. install fixtures

3. What type of faucet is sometimes used in public buildings specifically to conserve water?

 a. Self-closing faucet

 b. Compression faucet

 c. Quick pression faucet

 d. Quick opening faucet

4. Which of the following slopes is the most common for main house drains?

 a. $\frac{1}{16}$" per foot

 b. $\frac{1}{8}$" per foot

 c. $\frac{1}{4}$" per foot

 d. $\frac{1}{2}$" per foot

Study Tip

Water closet compartments shall have a minimum dimension of 30" × 60".

Study Tip

A septic tank must be a minimum of 25 feet away from a private water supply.

5. A magnesium rod used to retard corrosion in a water heater is called a _____.

 a. dip tube

 b. stabilizer rod

 c. adapter tube

 d. sacrificial rod

6. A relief valve is used to prevent damage from excessive pressure in a closed water, air, or steam system. A small relief valve, ½" to 2", is made of _____.

 a. brass with threaded connections

 b. cast iron

 c. steel with brass seats

 d. All of the above

7. Which of the following is true about a water-filled trap?

 a. Siphonage will upset trap seal

 b. Backpressure will upset trap seal

 c. Both a and b

 d. Neither a nor b

8. What prevents waste from slowing down at the base of a stack and forming a slug of water that would cause pressure changes?

 a. Aerator

 b. Deaerator

 c. Separation chamber

 d. None of the above

9. Which type of drain fitting is commonly installed on bathtubs?

 a. Waste fitting

 b. Overflow fitting

 c. Diverter fitting

 d. Lift waste fitting

10. Which of the following fixtures is not considered when if installed with 25' of vented drainage pipeline?

 a. Lavatory

 b. Bidet

 c. Water closet

 d. Floor drain

11. Which of the following is not true when considering minimum requirements of a floor drain?

 a. Floor drain strainer must be removable

 b. Minimum outlet size of a floor drain is $1\frac{1}{2}$"

 c. The combined free area of the holes in the grate must equal the drain outlet size

 d. Must be equipped with a backwater valve when installed below ground level

12. All of the following are necessary when installing a drinking fountain that discharges indirectly into the drainage system except _____.

 a. $1\frac{1}{4}$" vent pipe

 b. $1\frac{1}{4}$" P-trap

 c. $1\frac{1}{4}$" waste pipe

 d. All of the above

Study Tip

A compression faucet uses a rubber-like washer compressed against metal.

Study Tip

Aeration systems are used when it is impossible to install a leach field.

13. Scaled views that show the shape and size of the building lot and the location, shape, and overall size of a house are known as _____.

 a. plot plans

 b. details

 c. sectional views

 d. floor plans

14. Which of the following results from negative pressure in the distribution pipes of a potable water supply system?

 a. Backpressure

 b. Backsiphonage

 c. Flooding

 d. None of the above

15. Convert the following into feet: 10 feet 10 inches

 a. 10.10'

 b. 10.12'

 c. 10.83'

 d. 130"

16. What is the capacity, in gallons, of a 2-inch pipe that measures 10 feet long? Given: Volume = $0.0408 \times D, L$

 a. 0.82 gal.

 b. 1 gal.

 c. 1.25 gal.

 d. 1.63 gal.

17. Horizontal drainage pipeline is designed to flow about _____ full.

 a. ⅓

 b. ⅔

 c. ¾

 d. ½

18. Which type of fitting has the same dimensions at all openings?

 a. Straight

 b. Standard

 c. Reducing

 d. None of the above

19. Which of the following is not true about gas piping?

 a. Aluminum pipe will not be used outside or underground

 b. Plastic pipe may only be used inside

 c. Aluminum pipe must be factory coated where it may be in contact with detergent

 d. All of the above

20. Septic tanks should be pumped out every _____ years.

 a. 1 to 2

 b. 3 to 5

 c. 5 to 7

 d. 8 to 10

21. All porcelain enamel surfaces on plumbing fixtures shall be _____.

 a. shatter resistant

 b. acid resistant

 c. heat resistant

 d. freeze resistant

Study Tip

The minimum size stack vent shall be no less than ½ the diameter of the drain.

Study Tip

Volume of a cylinder is calculated by multiplying $\pi r^2 h$.

22. Which of the following is considered to be hazardous material?

 a. Thread cutting oil

 b. Cleaning solvent

 c. Solvent cement

 d. All of the above

23. Decomposition of organic material found in a sewer is a hazardous by-product called _____.

 a. sludge

 b. soil

 c. sewer gas

 d. oxygen

24. A(n) _____ plug is a test device installed on the inside of the pipe end.

 a. inspection

 b. pipe

 c. gauge

 d. inlet

25. A(n) _____ cap is a test device installed on the outside of a pipe end.

 a. pipe

 b. gauge

 c. tee

 d. outside

"IDENTIFY" QUIZ

1. Identify this tool.

 a. End pipe wrench

 b. Straight pipe wrench

 c. Monkey wrench

 d. Adjustable wrench

Study Tip

The most likely water supply pipe to leak due to deterioration is galvanized pipe.

2. Identify the valve.

 a. Gate

 b. Globe

 c. Angle

 d. Rising stem

3. Identify this tool.

 a. Graphite placement device

 b. Pen

 c. Pencil

 d. Soap stone

4. Identify this fixture.

 a. Bottom outlet floor drain

 b. Bottom outlet roof drain

 c. Side outlet floor drain

 d. Side outlet roof drain

5. Identify this tool.

 a. Squeeze type cutter

 b. Chain ratchet cutter

 c. Hydraulic cutter

 d. Hand cutter

Study Tip

Sludge is the decomposition of organic material found at the bottom of a septic tank.

6. Identify this tool.

 a. Pipe reamer

 b. Pipe cutter

 c. Pipe threader

 d. Pipe stand

7. Identify the fitting.

 a. Male adaptor

 b. Plug

 c. Reducer

 d. Union

8. Identify this fitting.

 a. Copper to FIP adapter

 b. Copper coupling

 c. Copper union

 d. Copper to MIP adapter

9. Identify this fitting.

 a. Vent tee

 b. Sanitary tee

 c. Sweep tee

 d. Cleanout tee

10. Identify this tool.

 a. Universal saw

 b. Ratchet cutter

 c. Tubing cutter

 d. Crimping tool

11. Identify the fitting.

 a. Bell and spigot tee

 b. Solvent welded tee

 c. Screwed tee

 d. Insert tee

Study Tip

The chain wrench is used on large diameter of pipe.

12. Identify this tool.

 a. End pipe wrench

 b. Strap pipe wrench

 c. Compound wrench

 d. Chain wrench

13. Identify the fitting.

 a. Bell and spigot cross

 b. Double sanitary
 tee—no hub

 c. Plastic cross

 d. PVC cross

14. Identify this tool.

 a. Smooth-jawed pliers

 b. Spud wrench

 c. Offset hex wrench

 d. Adjustable wrench

15. Identify this tool.

 a. Tub wrench

 b. Sink wrench

 c. Basket strainer wrench

 d. Basin wrench

Study Tip

A pipe wrench has heavy toothed jaws for gripping pipe.

16. Identify this tool.

 a. Monkey wrench

 b. Tub wrench

 c. Internal wrench

 d. Offset hex wrench

17. Identify this tool.

 a. Flaring tool

 b. Adjustable wrench

 c. Tube cutter

 d. Pipe reamer

18. Identify this tool.

 a. Inner–outer copper cleaning tool

 b. Copper pipe reamer

 c. Copper tube cutter

 d. Copper fitting brush

19. Identify this tool.

 a. Pipe wrench

 b. Crescent wrench

 c. Strap wrench

 d. Chain wrench

20. Identify this tool.

 a. Spud wrench

 b. Sink drain wrench

 c. Garbage disposal wrench

 d. Internal wrench

21. Identify this tool.

 a. Delta faucet repair wrench

 b. Garbage disposal wrench

 c. Allen wrench

 d. Faucet seat removal wrench

22. Identify this tool.

 a. Flushometer valve wrench

 b. Fit-all wrench

 c. PO plug wrench

 d. Check valve repair wrench

23. Identify this tool.

 a. Smooth-jaw pliers

 b. Monkey wrench

 c. Water pump pliers

 d. Slip-joint pliers

24. Identify this tool.

 a. Sink basket strainer wrench

 b. PO plug wrench

 c. Tub drain wrench

 d. Fit-all wrench

25. Identify this tool.

 a. Faucet handle puller

 b. Seal puller

 c. Moen cartridge puller

 d. None of the above

Study Tip

The basin wrench fits in narrow recesses to turn water supply nuts.

Notes

PART EIGHT
Project Management

The following questions are typical of those found in the Project Management section of most state contractor licensing exams. While many state requirements include separate exams to cover this content, several states incorporate these questions intermittently with the trade questions. Be careful to read your exam candidate bulletin to determine whether your exam covers this subject matter and to determine the approved reference material.

A monkey wrench has one movable jaw. The jaws provide no clamping action when force is applied.

PROJECT MANAGEMENT PRACTICE EXAM #1

1. Each of the following is regulated by the Fair Labor Standards Act EXCEPT:

 a. Vacation

 b. Time and a half for hours over 40

 c. Minimum wage

 d. Compliance with child labor laws

2. What amount does an employer pay for Social Security and Medicare?

 a. 6.2%

 b. 1.45%

 c. 7.65%

 d. 5.4% on the first $7,000

3. Two brothers are planning to start a plumbing contracting company together. The oldest brother will be the financial backer primarily and wants no day-to-day involvement in the operations. What type of business should the brothers form?

 a. General partnership

 b. Proprietorship

 c. Limited partnership

 d. Corporation

4. The best source for projecting future project and company overhead percentage rates is:

 a. Historical data

 b. Bid documents

 c. Unit pricing

 d. Job cost records

5. Which of the following insurance types is most widely used for construction projects?

 a. Building operations

 b. Liability

 c. All-risk builder's risk

 d. Catastrophe

6. Which of the following would be considered project overhead?

 a. Yellow page advertising

 b. Secretary's salary

 c. Electric bill for the office

 d. Electric bill for project office

7. Yellow Duck Plumbing Company has 21 employees. According to OSHA, this company must keep a record for _____ years after each recorded accident for each one of its employees while on the job site.

 a. 3

 b. 4

 c. 5

 d. 6

8. What entity continues to exist beyond the death of one of the principals?

 a. Corporation

 b. Partnership

 c. Sole proprietorship

 d. Limited partnership

Study Tip

Corporations run the risk of double taxation, not only on the profits but on the dividends as well.

9. Employers are not allowed to deduct _____ from employee's wages.

 a. Social Security

 b. Medicare

 c. Federal Unemployment Tax (FUTA)

 d. Federal Income Tax

10. Which of the following is a disadvantage of setting up a general partnership?

 a. Single taxation

 b. Easy setup

 c. Unlimited personal liability

 d. The inability to pool financial resources

11. Which of the following would NOT constitute a material breach of contract?

 a. Building on the wrong site

 b. A refusal to work

 c. The owner refuses to pay

 d. A delay in the work

12. What form of business organization offers contractors the most protection against personal liability?

 a. Corporation

 b. Partnership

 c. Joint venture

 d. Sole proprietorship

13. <u>Given:</u> A carpenter drops a hammer from a roof and injures a curious bystander. Which type of insurance would cover the victim's medical costs?

 a. Builder's risk

 b. Named-peril risk

 c. Liability

 d. Risk management

14. The simplest form of business ownership is:

 a. Corporation

 b. Partnership

 c. Limited partnership

 d. Sole proprietorship

15. What form is used to report Federal Unemployment taxes (FUTA)?

 a. W-3

 b. 940

 c. W-2

 d. 1099

16. A large plumbing contractor maintains a central office and has crews working at several projects. Which of the following is considered a company overhead cost?

 a. Superintendent salary

 b. Temporary barricades

 c. Site cleanup

 d. Advertising

Study Tip

An indirect cost of a company cannot be tied to a single project.

Study Tip

*Starting a corporation is forming
a new entity that protects
shareholders.*

17. Typically, the only party with whom the subcontractors and their employees have a contractual relationship with is the:

 a. Owner's representative

 b. Architect

 c. Owner

 d. Contractor

18. The purpose of the journal:

 a. To post transactions on a weekly basis

 b. To post transactions on a daily basis

 c. To ensure proper job costing

 d. To balance the checkbook

19. What is the latest date an employer can file the Federal Unemployment Tax Return (Form 940)? (Assume no time extensions are taken and the balance due is greater than $100.00)

 a. December 1

 b. December 31

 c. January 1

 d. January 31

20. Employee records regarding pay shall be kept for at least _____ years.

 a. 1

 b. 2

 c. 3

 d. 4

21. Which of the following is the proper formula for determining ASSETS?

 a. Liabilities + Owner's equity

 b. Liabilities − Owner's equity

 c. Owner's equity − Liabilities

 d. Assets − Liabilities

22. Which of the following components make a change order complete?

 a. The change to be made, the charge, and the effect on time

 b. The owner's/contractor's address and the change order number

 c. The legal description and the contractor's license number

 d. The contractor's license number, the change order number, and the architect's signature

Study Tip

Employer's liability insurance protects employees over above worker's compensation due to an injury or accident.

23. The primary purpose of a written contract is to:

 a. Define rights and obligations of each party

 b. Standardize all requirements of work performance

 c. Condense construction requirements

 d. Standardize all requirements of work performance

24. The term used in scheduling that makes reference to the TIME a project can be delayed without extending the project completion:

 a. Float time

 b. Lag time

 c. Critical task

 d. Non-critical task

25. Which of the following terms is used to describe withholding a percentage of a progress payment pending completion to a certain point for the purpose of covering defects or faulty installation?

 a. Retainage

 b. Withholding

 c. Holdback

 d. None of the above

Study Tip

The accrual method of accounting records income at the time it is earned even though payment may not have been rendered.

PROJECT MANAGEMENT PRACTICE EXAM #2

1. A bid bond provides _____.

 a. assurance that the bonding company is liable for all bid discrepancies

 b. assurance to the owner or government agency that the contractor will honor the bid submitted by entering into a contract in accordance with the terms of the bid

 c. a guarantee to the owner that the project will be completed

 d. a guarantee of both job completion and payment of all labor and materials

2. A truck that is due to be paid off within a year would be considered a(n):

 a. Current liability

 b. Long-term liability

 c. Asset

 d. Other

3. In the event a subcontractor does not carry worker's compensation insurance on his employees, who is responsible for providing the insurance?

 a. Owner

 b. General contractor

 c. State government

 d. No one

4. Which of the following financial statements indicate the financial condition of a company at a given point in time?

 a. Income statement

 b. Cash flow statement

 c. Balance sheet

 d. Funds flow statement

5. A first aid kit is checked before it is sent to the job, and then checked _____.

 a. Every week

 b. Bi-weekly

 c. Monthly

 d. Quarterly

Study Tip

Two methods of scheduling are the critical path and the bar chart.

6. Which of the following would be classified as a long-term debt?

 a. Current annuity of long-term debt

 b. Mortgage on company office building

 c. Accounts payable

 d. Accumulated depreciation

7. What is another name for owner's equity?

 a. Operating capital

 b. Liquid assets

 c. Net worth

 d. Liability

8. Which of the following statements defines *Networking* Capital?

 a. The amount the current assets exceeds the current liabilities

 b. The amount listed as an item overhead expense

 c. The difference between the direct cost and the cost of sales

 d. Current liabilities less overhead expense

9. If two companies enter into a project as a joint venture, if one company does not have a contractor's license, what must occur?

 a. Company with license will manage

 b. A license in the name of the joint venture shall be obtained

 c. This is permitted as long as at least one firm has a license

 d. Other

Study Tip

Performance bond guarantees the project will be completed according to plans and specifications.

10. Based on an accrual basis of accounting, earnings, which have not been received, will be displayed as _____.

 a. current assets

 b. fixed assets

 c. current liabilities

 d. owner's equities

11. The required quality of a project is described in the:

 a. Budget

 b. Specifications

 c. Plans

 d General conditions

12. Current assets are assets that can be converted into cash within _____.

 a. 1 year

 b. 6 months

 c. 1 month

 d. 2 years

13. A form of business organization that has a tax structure similar to a partnership but is organized as a corporation:

 a. Limited partnership

 b. Subchapter S corporation

 c. General partnership

 d. Non-profit organization

14. Regarding contracts, which of the following statements is NOT true?

 a. Words prevail over numbers

 b. General clauses prevail over special clauses

 c. When properly documented, handwritten agreements prevail over typed provisions.

 d. A contract deemed ambiguous is interpreted against the party that provided it.

Study Tip

Net worth = Assets − Liabilities.

15. Which of the following work-related injuries does NOT require the employer to report to OSHA?

 a. A broken wrist

 b. An X-ray

 c. Cancer due as a result of long-term exposure to work-related chemicals

 d. A sprain that removed an employee from work for 2 days

16. What is a benefit of good inventory control?

 a. Better for safety

 b. Better preparation for increase in business

 c. Reduced time and expense in handling materials

 d. Ability to ship faster to offsite storage

17. Periodic payments made by the owner to the contractor over a long project are called:

 a. Progress payments

 b. Acceleration payments

 c. Performance payments

 d. Provisional payments

Study Tip

A safety lanyard is required when 6' or more above the ground.

18. What accounting method records income when cash is received?

 a. Accrual basis

 b. Cash basis

 c. Single entry

 d. Double entry

19. What portion of the worker's compensation insurance premium can an employer charge to an employee?

 a. A maximum of 2% of the employee's weekly gross earnings

 b. No deduction is allowed

 c. Up to 50% of the premium

 d. A flat fee of $25 per week

20. If an employee is employed by a corporation and takes the licensing examination for the corporation, who will own the license when it is issued?

 a. The employee

 b. The stockholders

 c. The president of the corporation

 d. The corporation itself

21. An employer must maintain records about work-related injuries for _____.

 a. 3 years

 b. 5 years

 c. 1 year

 d. 7 years

22. Which of the following statements is NOT true regarding a bar chart method of scheduling?

 a. Shows the start and finish times for each task

 b. Does not show how tasks relate to each other

 c. Most common scheduling system found on the job site

 d. None of the above (all are true)

23. When a clause in a contract calls for liquidation damages, it is referring to:

 a. A payment per day for finishing after contract deadline

 b. A payment due to flooding

 c. An expedient payment for improper work

 d. None of the above

24. Which of the following statements is true about partnerships?

 a. Each partner may cause the other to be jointly liable for any action incurred during normal business

 b. Each partner is personally liable for his individual actions only

 c. Each partner may sell his share of the partnership without the consent of the other partners

 d. A general partner may discharge himself from all liability by declaring himself to be a limited partner

25. Which of the following acts requires payment of prevailing wage rates and fringe benefits on federally assisted construction?

 a. Davis-Bacon Act

 b. Walsh-Healey Public Contracts Act

 c. Wage Garnishment Law

 d. Fair Labor Standards Act

Study Tip

Retainage is typically 10% of each progress payment.

Study Tip

Time cards must stay on file for at least 3 years.

USEFUL INFORMATION

The most common forms of business organizations include sole proprietorships, corporations, general partnerships, and limited partnerships. Business types are typically chosen with the advice of an attorney and/or accountant based on the degree of liability the owner is willing to accept. In other words, the trade-off for less liability is more regulation and a greater tax burden.

Costs that relate to the construction activity itself but are not easily identifiable costs of a given project are referred to as project overhead. Company overhead, on the other hand, is costs that continue even when no work is under way. Examples of company overhead include office salaries, office rent, company insurance, and accounting fees. Examples of project overhead include a supervisor's salary, permit fees, bonds, and temporary toilets.

For a contract to be valid, there must be a mutual understanding; it must contain an offer, an acceptance, and a consideration. If the initial offer is not accepted, a counteroffer is made. If the offer is accepted, it must be followed with consideration which is usually in the form of money. Many times negotiation precludes the offer. The difference between an offer and a negotiation is that an offer is specific with price and completion dates. Mere negotiation is much less specific.

A bond is a three-party agreement between the insurance/surety company, the contractor, and the owner (contractor's client). Upon issuance of the bond, the surety company assumes the liability for the contractor. The only party able to claim against a bond purchased by a contractor is the owner. Many different types of bonds exist such as bid bond, performance bond, and payment bond. In the private sector of commercial construction, it is very common for the contractor to require the subcontractor to purchase a bond. The Miller Act requires bonds on federal construction projects.

Each employer with 11 or more employees is required to maintain records of all occupational injuries and illnesses regardless of their severity. Employers with 10 or fewer employees enjoy less stringent requirements by OSHA regarding reporting and record keeping; however, all employers must report fatalities and incidents resulting in multiple hospitalization. If a business changes ownership, an owner is responsible for recording and reporting work-related injuries and illnesses only for that period of the year during which he owned the establishment.

The Fair Labor Standards Act requires payment of federal minimum wage and overtime payment of at least one and one-half the employee's regular rate of pay for hours worked over 40 during a fixed and recurring 168 consecutive hour workweek. Many times, supervisors and members of management are exempt from many of these requirements. Employers should be careful to understand that a salaried employee does not always mean the FLSA rules of overtime will not apply.

To prove both employment eligibility and identity, a Form I-9 must be completed within 3 days of hiring a new employee. This form serves as documentation that the employer has obtained the necessary documents for verification. In the event a driver's license is provided to prove identity, a social security card must be provided to prove employment eligibility. The I-9 form must be kept on file for at least 3 years from the date of employment or for 1 year after the employee leaves the job, whichever is later.

There are two basic types of accounting: accrual basis and cash basis. Under the cash basis of accounting, income is recorded when it is received and expenses are recorded when payment is made. In contrast, under the accrual basis of accounting, income is recognized when it is earned regardless of when it is received and expenses are recorded when incurred regardless of when they are paid.

A qualifying agent is a full-time employee or officer of a company that has either volunteered or been selected to be the test-taker for the entity. Typically, the company's financials, rather than the qualifying agent, is presented to the appropriate agency for licensing consideration. When a qualifying agent terminates the relationship, the license will remain the property of the company. The company has a specified amount of time to replace the agent.

Study Tip

Wage garnishment law prohibits the firing of an individual due to a single wage attachment.

Notes

PART NINE
OSHA

The following section is based on OSHA 1926, Construction Industry Regulations. Generally, only a small portion of plumbing exams pertains to this subject matter. If the OSHA book is an approved reference material listed for your exam, these questions represent the types of questions that will be covered on your exam.

OSHA PRACTICE EXAM

1. The maximum slope for an 8' excavation in soil type B:

 a. ¾:1

 b. ½:1

 c. 1:1

 d. 2:1

2. The minimum perpendicular clearance between fixed ladder rungs, cleats, steps and any obstruction behind the ladder is:

 a. 6"

 b. 7"

 c. 8"

 d. 9"

3. The use of electrical cords for hoisting or lowering shall _____.

 a. be allowed

 b. not be allowed

 c. only be permitted when load is less than 20 lbs.

 d. only be permitted when load is less than 30 lbs.

4. When necessary to require employees to work in a trench, the excavated materials shall be placed a minimum of ____ feet away from the edge of the trench.

 a. 0

 b. 1

 c. 2

 d. 3

5. Eye protection near dangerous working conditions _____.

 a. is required at the employee's expense

 b. is required at the employer's expense

 c. is not required, except by union regulations

 d. is not required

6. Who is responsible for the fire safety and prevention on a job site through completion of the project?

 a. OSHA

 b. Employer

 c. Fire marshall

 d. Building inspector

7. When employees are required to be in trenches 4 feet deep or more, an adequate means of exit such as a ladder or steps shall be provided and located so as to require no more than ___ feet of lateral travel.

 a. 10

 b. 15

 c. 25

 d. 30

8. How much water is allowed in an excavation before work must cease:

 a. 3" or less at edge

 b. 2" or less at edge

 c. 1" or less at edge

 d. None

Notes

Study Tip

OSHA must be notified within 8 hours if there is a fatality or multiple hospitalization resulting from an accident on a job site.

9. The minimum illumination in active general construction storage areas is ____.

 a. 3-foot candles

 b. 5-foot candles

 c. 10-foot candles

 d. 30-foot candles

10. After being sent to the job site, a first aid kit is checked at least _____ .

 a. weekly

 b. bi-weekly

 c. monthly

 d. quarterly

11. OSHA requires a ladder in every excavation in excess of:

 a. 5'

 b. 3'

 c. 4'

 d. 6'

12. Compressed air shall not be used for cleaning purposes, except where reduced to less than _____ psi.

 a. 10

 b. 20

 c. 30

 d. 70

13. Sand is considered what soil type?

 a. A

 b. B

 c. C

 d. D

14. According to OSHA, when employees are exposed to inhalation, ingestion, skin absorption, or contact with any substance at concentration above the _____ they shall be evacuated unless specific forms of personnel protective equipment or devices are provided.

 a. threshold limit value

 b. PPM mandate

 c. PEL levels

 d. oxygen deprivation level

15. According to OSHA the height at which no fall protection is required is ____ feet.

 a. 6

 b. 8

 c. 10

 d. 12

16. Circular saws shall be equipped with a _____ switch.

 a. position on–off

 b. momentary on–off

 c. lock on control

 d. constant pressure switch

17. Combustible debris should be removed from a construction job site _____.

 a. when the job is complete

 b. only in the event of hazardous fire conditions

 c. periodically by any means

 d. at regular intervals

Study Tip

Hardhats are provided by the employer.

Study Tip

Dielectric unions prevent galvanic action.

18. When using a stepladder:

 a. The top step can be used

 b. The cross-bracing on the back can be used as a step

 c. Both hands must be used on the sides

 d. The ladder should be inspected by a competent person for visible defects on a periodic basis

19. When employees are required to be in trenches 4 feet or more, an adequate means of egress shall be provided and located so as to require not more lateral travel than _____.

 a. 10'

 b. 12'

 c. 15'

 d. 25'

20. The maximum allowable slope for excavations less than 20 feet in depth for type C soil is _____.

 a. ¾ horizontal to 1 vertical

 b. 1 horizontal to 1 vertical

 c. 1½ horizontal to 1 vertical

 d. 2 horizontal to 1 vertical

21. A 10-foot-deep trench is to be dug in granular sand. What is the max allowable bank slope?

 a. 1:1

 b. 1½:1

 c. ½:1

 d. ¼:1

22. When using a portable ladder for reaching an elevated surface or for ingress or egress of an excavation, the ladder must extend what minimum height above the surface?

 a. 2.5'

 b. 3'

 c. 3.5'

 d. 4'

23. _____ soil has a maximum allowable slope of 1:1.

 a. Type A

 b. Type B

 c. Type C

 d. Type D

24. Each of the following is a soil type EXCEPT:

 a. Gravel

 b. Sediment

 c. Sand

 d. Clay

25. OSHA requires the employer to provide _____ for all employees/ workers.

 a. no smoking areas

 b. a safe working environment

 c. new tools

 d. hardhats

Study Tip

Combustible debris could be dust or fibers.

Notes

PART TEN

Backflow Prevention

BACKFLOW PREVENTION PRACTICE EXAM

1. How can a backsiphonage action be broken?

 a. Air gap

 b. DCVA

 c. RPBP

 d. Water purveyor

2. Atmospheric pressure at sea level is _____ psi.

 a. 33.9

 b. 21.7

 c. 17.4

 d. 14.7

3. If a column of water is exerting a pressure of 50 psi, how high must the water be inside the column?

 a. 27.7 feet

 b. 115.5 feet

 c. 132.5 feet

 d. 150.0 feet

4. Reversal in the normal direction of flow is called:

 a. Backflow

 b. Backsiphonage

 c. Cross-connection

 d. Hazardous

5. Which of the following assemblies can only be used for non-hazardous application?

 a. Air gap

 b. DCVA

 c. PVB

 d. RPBP

6. At what point is the relief valve in an RP required to open?

 a. 1.0 psid

 b. 2.0 psid

 c. 3.0 psid

 d. 5.0 psid

7. According to USC, what is the correct height for installing an RPZ?

 a. 6 inches above grade

 b. 12 inches above grade, but no more than 24 inches

 c. 12 inches above grade, but no more than 36 inches

 d. Does not matter as long as it can be tested

8. While testing check valve #1 on a DC, water recedes from the tube attached to test cock #3. The gage should be moved to the centerline of the assembly and the static pressure drop across the No. 1 check should be recorded. What other information should be recorded?

 a. Check valve #2 leaking

 b. Shutoff valve #2 leaking

 c. Both a and b

 d. No additional information is needed

9. In a perfect vacuum, water can be drawn up in a column _____ feet.

 a. 43.3

 b. 33.9

 c. 14.7

 d. 2.31

Notes

10. Which of the following is considered to be a biological health hazard?

 a. Yeast

 b. Giardia

 c. Fluoride

 d. Cyanide

11. Which of the following is the most suitable protection when directly connecting potable water supply lines to sewerage systems or to radioactive sources?

 a. Air gap

 b. DCVA

 c. PVB

 d. RPBP

12. There are two types of backflow conditions. What are they?

 a. Backpressure and backsiphonage

 b. Backsiphonage and capillary action

 c. Isolation and containment

 d. Potable and non-potable

13. Which of the following is considered to be a chemical health hazard?

 a. Arsenic

 b. Cryptosporidium

 c. Iron

 d. Legion Ella

14. What is the correct height to install a PVB?

 a. 12 inches above the ground

 b. 12 inches above the highest downstream point

 c. 6 inches above the ground

 d. 6 inches above the highest downstream point

15. What is the minimum spring tension recommended on the #1 check valve in an RPZ?

 a. 5 psi

 b. 3 psi

 c. 3 psi greater than the tension on the relief valve

 d. 2 psi greater than the tension on the relief valve

16. Which of the following is NOT a criterion for an approved assembly?

 a. Shutoff valves before and after backflow device

 b. Shutoff valves have full flow and be resilient

 c. Shipped by manufacturer in separate parts for inspection

 d. Accommodates testing and repairs in-line

17. When testing a double check valve assembly, what is the correct order for closing the shutoff valves?

 a. Shutoff #1, then shutoff #2

 b. Shutoff #2, then shutoff #1

 c. It does not matter which shutoff valve is closed first

 d. Only the #2 shutoff valve is used in testing this device

Notes

18. What is the minimum period of trouble-free field service required for a device to gain approval by the Foundation for Cross-Connection Control and Hydraulic Research at USC?

 a. 3 years

 b. 6 months

 c. 12 months

 d. 2 years

19. What Federal Law made water purveyors responsible for water quality all the way into consumers' homes or businesses?

 a. Clean Water Act of 1

 b. Safe Drinking Water Act Amendment of 1986

 c. Surface Water Treatment Rule

 d. Water Preservation Act of 1974

20. What is the minimum pressure drop across a spring-loaded, soft-seated check valve?

 a. 1.0 psid

 b. 2.0 psid

 c. 3.0 psid

 d. 5.0 psid

21. What type of backflow prevention assembly would be recommended in a car wash facility at the service connection to a potable water supply?

 a. AVB

 b. DCVA

 c. PVB

 d. RPZ

22. The condition where undersized water pipes carry water under high velocity and may cause a backsiphonage is referred to as the

 a. Darwin Award

 b. Doppler Shift Effect

 c. Bernoulli Law

 d. Venturi Effect

23. When should an RP device be tested?

 a. At least annually

 b. After installation

 c. After repairs

 d. All of the above

24. What protection is equivalent to or better than an RP?

 a. Approved air-gap separation

 b. Atmospheric vacuum breaker

 c. Pressure vacuum breaker

 d. Double check valve assembly

25. What type of gauge is used to measure the pressure across check valves in RPs, DCs, and PVBs?

 a. Differential

 b. Hydrant

 c. Pitot

 d. Venturi

26. When testing a reduced pressure zone backflow preventer, bleed valves should be **opened** in which order?

 a. High then low

 b. Low then high

 c. Simultaneously

 d. It does not matter in which order they are opened

Notes

27. While testing a Reduce Pressure Backflow Preventer, the tester has determined that the line pressure is 75 psi. The test shows that check valve #1 holds a pressure differential of 7.5, check valve #2 holds a pressure differential of 2.2, and the relief opened at 1.8 psid. What is the line pressure downstream of the RP in a no-flow condition?

 a. 9.7

 b. 11.5

 c. 63.5

 d. 65.3

28. A link between a potable water supply and some other non-potable supply or source is called a:

 a. Backflow

 b. Cross-loop

 c. Cross-connection

 d. Hazardous connection

29. What single difficult-to-regulate item is most commonly associated with cross-connections, specifically through backsiphonage?

 a. Aspirator

 b. Faulty valves

 c. Fire protection systems

 d. Garden hose

30. The appropriate height for an air gap is:

 a. Twice the pipe diameter above the flood rim

 b. Twice the diameter of the pipe above the flood rim of vessel, minimum of 1 inch

 c. 1 inch above the flood rim plus the diameter of the pipe squared

 d. 1 inch

31. Backflow prevention assemblies are often installed in parallel in order to:

 a. Allow more water to pass through during peak demands

 b. Comply with Federal and State standards

 c. Maintain service during testing and repairs

 d. Minimize usage for longer life of assemblies

32. Where is the relief valve of RPZ located?

 a. Between test cock #1 and check valve #1

 b. Between check valve #1 and check valve #2

 c. Between check valve #2 and test cock #3

 d. Immediately after test cock #3

33. According to the Safe Drinking Act, the primary responsibility for preventing water from an unsuitable or unsafe source from entering the public water system lies with the:

 a. Consumer who is connected to the system

 b. Health agency

 c. Plumber

 d. Water utility supplying the water

34. Why do you open and then close the "low" bleed in an RP device before determining the pressure differential across check valve #1?

 a. To check #1 check valve for leakage

 b. To create a backsiphonage situation

 c. To determine the relief valve opening point

 d. Reestablish "setup" pressure

Notes

35. Static pressure means:

 a. Electricity generated by flowing water across a dielectric surface

 b. Flow

 c. No flow

 d. The difference between the higher pressure upstream and the lower pressure downstream

36. An atmospheric vacuum breaker is designed to protect against a non-health hazard (pollutant) or a health hazard (contaminant) under what condition or conditions?

 a. Backpressure

 b. Backsiphonage only

 c. Both backpressure and backsiphonage

 d. None of the above

37. Backflow prevention by containment means:

 a. Placing a backflow preventer at every connection

 b. Placing a backflow preventer at the meter

 c. Placing a backflow preventer at the boiler

 d. Placing a backflow preventer at the beginning of the sewer

38. Should a customer refuse to install a backflow prevention device at a service connection where a potential health hazard exists, the water purveyor should have the authority to:

 a. Discontinue water service

 b. Install the required backflow prevention assembly free of charge

 c. Refer the situation to the health department

 d. Take legal action

39. What is the purpose of the 3.0-psi buffer in an RPZ when check valves in other assemblies only require a 1.0-psi static pressure drop across them?

 a. To minimize the effects of small fluctuations in line pressure so as not to cause the differential pressure relief valve to discharge

 b. Line pressure through an RPZ is much greater than that through other types of assemblies which require the greater check valve tensions

 c. An RPZ which maintains a buffer less than 3.0 psi cannot provide adequate backflow protection

 d. To permit the relief valve to discharge when the assembly is placed under backpressure conditions

40. The best place to install a reduced pressure backflow prevention device is:

 a. At or near the ceiling in the basement

 b. In an accessible location above grade

 c. In a pit, provided it has a cover

 d. Underground

41. Which of the following markings is NOT required for backflow prevention assemblies?

 a. Direction of flow

 b. Maximum rated working water pressure and water temperature

 c. Minimum working water pressure for proper operation

 d. Size and model number

42. What are the limitations of an air-gap separation?

 a. Water pressure is lost

 b. Only acceptable for low hazards

 c. Easily altered

 d. a and c only

 e. All of a, b, and c

43. As the velocity of water increases, pressure:

 a. Increases proportionally

 b. Increases at twice the rate of velocity increase

 c. Decreases

 d. Remains constant

44. What company or agency is responsible for testing and approving all backflow prevention assemblies in the United States?

 a. American Backflow Prevention Association

 b. American Water Works Association

 c. Environmental Protection Agency

 d. Foundation for Cross-Connection Control and Hydraulic Research

45. What are the conditions checked when testing a reduced pressure backflow preventer?

 a. Test #1 check valve for leakage

 b. Test #2 check valve for leakage

 c. Test the relief valve to see if it will open at a differential of 2 psid or greater

 d. All of the above

46. When setting up an RP for testing you notice that the gauge dropped to the low end of the scale and that the relief valve began discharging continuously when the shutoff valve #2 was closed. What is the probable cause?

 a. Leaking shutoff valve #2

 b. Check valve #2 failed under backpressure

 c. Leaking check valve #1

 d. Relief valve leaks

47. If you observe the relief valve of an RPZ "spitting" water, what would be the suspected cause?

 a. Relief valve leaks

 b. Check valve #2 is leaking

 c. Check valve #1 is leaking

 d. Pressure fluctuating

48. You approach an RP that is dripping and the customer is currently using water. You close shutoff valve #2 and the relief valve continues to leak. What does this indicate?

 a. Relief valve leaks

 b. Leaking check valve #2

 c. Leaking check valve #1

 d. Pressure fluctuating

49. You approach an RP that is dripping. You first move to close shutoff valve #2 at which time the relief valve stops dripping. What does this indicate?

 a. Relief valve leaks

 b. Check valve #2 is leaking under backpressure

 c. Check valve #1 is leaking

 d. Pressure fluctuating

50. The difference between the opening point of the relief valve and the differential across check valve #1 should be at least _____ psid to minimize excessive discharges due to pressure fluctuations in the supply line.

 a. five

 b. two

 c. one

 d. three

Notes

51. A vacuum breaker should be used to protect against
 _____.

 a. backpressure only

 b. backsiphonage only

 c. backpressure and backsiphonage

 d. leakage

52. An approved air gap shall be at least _____ the diameter of
 the supply pipe.

 a. one time

 b. three times

 c. four times

 d. twice

53. The specifications of the 9th Edition of the Manual for Cross
 Connection Control calls for the relief valve to be fully open
 when the pressure differential between the "zone" and the
 upstream line pressure is _____ psid or less.

 a. one

 b. two

 c. three

 d. zero

54. The term "auxiliary supply" means any water supply on or avail-
 able to the premises other than the _____ water supply.

 a. public

 b. cistern

 c. non-potable

 d. all the above

55. An appropriate application of a reduced pressure backflow preventer would be to provide adequate protection for what type or degree of hazard?

 a. Pollutant

 b. Virus

 c. High

 d. Low

56. For installations requiring continuous service and where it is not possible or practical to provide water service from two separate service lines entering the premises, it is sometimes necessary to install _____ devices.

 a. high-hazard

 b. low-hazard

 c. backsiphonage or backpressure

 d. dual of parallel

57. Regulations dealing with cross-connections require that the backflow prevention device be tested at least _____ a year.

 a. one time

 b. twice

 c. once

 d. both a and c

58. The minimum approved air gap is _____ inch(es).

 a. two

 b. one

 c. three

 d. there is no minimum

59. What are the two backflow conditions that may cause a reversal of flow in water lines?

 a. High and low hazards

 b. Ventura and aspirator

 c. Backpressure and backsiphonage

 d. Pollutants and contaminants

Answer the following questions based on the procedure to test the reduced pressure backflow preventer second check valve under reverse flow conditions.

Questions 60–65

60. High side hose should be connected to:

 a. Testcock #1

 b. Testcock #2

 c. Testcock #3

 d. Testcock #4

61. Low side hose should be connected to:

 a. Testcock #1

 b. Testcock #2

 c. Testcock #3

 d. Testcock #4

62. Bypass/vent hose should be connected to:

 a. Testcock #1

 b. Testcock #2

 c. Testcock #3

 d. Testcock #4

63. High side control valve should be:

 a. Closed

 b. Opened

64. Low side control valve should be:

 a. Closed

 b. Opened

65. Bypass/vent hose control valve should be:

 a. Closed

 b. Opened

66. A device consisting of a single check valve and an air opening. Device must be installed on the discharge side of the last valve.

 a. Atmospheric vacuum breaker

 b. Pressure vacuum breaker

 c. Spill-resistant vacuum breaker

 d. Double check valve assembly

67. A device having a spring leaded air opening downstream from a single check valve.

 a. Atmospheric vacuum breaker

 b. Pressure vacuum breaker

 c. Spill-resistant vacuum breaker

 d. Double check valve assembly

Notes

68. A form of backflow due to vacuum or partial vacuum.

 a. Barometer

 b. Cross-connection

 c. Backsiphonage

 d. Backpressure

69. A physical break between a supply pipe and a receiving vessel is a(n) _____.

 a. Cross-connection

 b. Air gap

 c. Backpressure

 d. Backsiphonage

70. Any actual or potential connection between a supply pipe and a public or private potable water system and any other source through which pollution may enter the potable system is

 _____.

 a. a Cross-connection

 b. an approved bypass

 c. permissible if OK'd by owner

 d. None of the above

71. A device consisting of two check valves and a differential relief valve located between the two check valves is called?

 a. Double check valve assembly

 b. Pressure vacuum breaker

 c. Reduced pressure principle assembly

 d. Spill-resistant vacuum breaker

72. A device consisting of two internally loaded check valves is called?

 a. Reduced pressure principle assembly

 b. Pressure vacuum breaker

 c. Atmospheric vacuum breaker

 d. Double check valve assembly

73. A form of backflow due to a higher pressure within a premise over supply pressure is called?

 a. Backpressure

 b. Backsiphonage

 c. Actual cross-connection

 d. Potential cross-connection

74. Which of the following is NOT a criterion for a Reduced Pressure Backflow Assembly device?

 a. Can be used for both high and low hazards.

 b. Has a relief valve located in the zone between the checks

 c. Cannot have any connections (outlets) between water meter and RP.

 d. Can be installed in a pit.

75. Which of the following is NOT a criterion for a double check valve device?

 a. Has two independently acting spring loaded check valves

 b. Can be used for both high and low hazards

 c. According to USC Foundation for Cross-Connection Control, must have a clearance of between 12 and 36 inches from bottom of device to ground/floor

 d. Must have 24-inch clearance of test cock side for maintenance

76. Which of the following is NOT a criterion for a pressure vacuum breaker device?

 a. Cannot have any valves or shutoffs on the discharge side of the device

 b. Has a spring-loaded check valve and a spring-loaded air inlet valve

 c. Must be installed at least 12 inches above receiving vessel or any outlet downstream of device

 d. Can be used for high and low hazard, backsiphonage only

PART ELEVEN
Answer Keys

Part 2: Math Practice Exam Answer Key

1. b

2. a Travel = Advance or offset \times 1.414; T = 28 \times 1.414 = 39.592 \approx 39½"

3. c Offset $=\sqrt{r^2 + s^2}$; 0 $= \sqrt{8^2 + 20^2}$; 0 $= \sqrt{64^2 + 400^2}$; 0 $= \sqrt{464}$; 0 = 21.54;
 Travel = 21.54 \times 1.414 = 30.45 \approx 30½ inches

4. a Fall = length \times % grade; F = 110' \times 0.04 = 4.4 feet \approx 4⅜ feet

5. d Pressure = 50'9" \times 0.434; P = 50.75 \times 0.434 = 22.03 psi

6. b Volume = 0.0408 $\times D^2 \times$ Length; V = 0.0408 $\times 6^2 \times$ 14;
 V = 0.0408 \times 36 \times 14 = 20.5632 gallons

7. c Head = Pressure \times 2.3; H = 40 \times 2.3 = 92 feet

8. d

9. d

10. b

11. b

12. a Fall = Length \times Pitch F = 75 \times ⅛ = 7⅝ = 9⅜ inches

13. d Pressure = Head \times 0.434 Pressure = 30 \times 0.434
 Pressure \approx 13 psi loss (Uphill)
 Pressure = 60 $-$ 13 = 47 psi

14. b

15. d

16. a Weight = 0.34 $\times D^2 \times L$; W = 0.34 $\times 10^2 \times$ 10
 $W - $ 0.34 \times 100 \times 10 = 340 pounds

17. d (volume in gallons = 0.0408 $\times d^2 \times L$(ft.)
 0.0408 \times 2304 \times 20 = 1880.06 gal.
 1880.06 \times 231 cu. in. = 434,294.78 cu. in.
 231 cu. in. = 1 gal.

18. b (cubic inches / 231 = gallons)
 434,294.78 cu. in. / 231 = 1880.06 gal.

19. c Travel = 1.414 \times advance T = 1.414 \times 48 =
 67.872 inches \approx 67⅞"
 67⅞" minus 3" (1½" for both fittings) = 64⅞" of pipe

20. a

21. d

22. b $4,450.00 \times 60% = $2,670.00

23. c Payment = Base \times Rate; Rate = Payment \div Base
 R = 900 \div 2195 = 0.41 = 41%

24. a

25. d

Notes

Part 3: Pipe, Fittings, and Symbols Questionnaire Answer Key

1. d
2. b
3. b
4. d
5. a
6. c
7. e
8. d
9. b
10. b
11. a
12. b
13. a
14. c
15. a
16. b
17. c
18. d
19. d
20. c
21. a
22. b
23. d
24. a
25. d

Part 4: IPC Practice Exam #1 Answer Key

1. b 305.1
2. a 305.6
3. d 402.4
4. b Table 308.5
5. d 419.2
6. b 412.3
7. c Table 303.4
8. d Tables 403.1 and 403.3
9. b 307.5
10. b 409.2
11. d 305.7
12. b Table 308.5
13. c 403.3.3
14. b 417.4
15. b 412.4
16. a 314.1
17. c 305.8
18. c Table 403.1
19. b 410.1
20. d 401.2
21. d 308.3
22. d 417.5
23. b 305.5
24. d 416.2
25. d 305.8

Notes

Part 4: IPC Practice Exam #2 Answer Key

1. d 602.3.1
2. d 504.7.2
3. c Table 604.3
4. a 603.2
5. b 606.5.4
6. b 608.3.1
7. a 501.6
8. d 504.4
9. c 602.3.5.1
10. a 605.24.1
11. d 605.2
12. c Table 608.1
13. a 608.8.1
14. b 502.3
15. d Table 604.5
16. a 505.1
17. b 603.1
18. b 605.20
19. b 606.2
20. d Table 604.4
21. c Table 605.3
22. c 606.5.7 (Table)
23. c Table 606.5.4
24. c 608.6.1
25. a Table 606.5.7

Notes

Part 4: IPC Practice Exam #3 Answer Key

1. b 701.7

2. d 708.2

3. d 713.9 and definition Chapter 2

4. b 711.2

5. c 713.11.4

6. d Table 710.1(1)

7. a 705.19.3

8. b Table 706.3 footnote

9. c Table 704.1

10. b Table 702.2

11. a 705.21

12. a 708.3.3

13. c 712.2

14. a Table 710.1(2)

15. b Table 712.4.2

16. a 712.3.2

17. c 705.14.2

18. b Table 709.1

19. b 708.3.2

20. d Table 709.2

21. c 708.7

22. d 709.3

23. a Table 709.1

24. a 708.8

25. b Table 709.1

Notes

Part 4: IPC Practice Exam #4 Answer Key

1. b Table 709.1

2. d Table 710.1(1) footnote: "Minimum size building drain serving a water closet shall be 3".

3. b Table 709.1 and Table 710.1(2) 2 + 2 + 2 = 6 dfu

4. b Table 709.1 6 sleeping rooms \times 6 dfu = 36

 Note: The question is not asking for total of all branch intervals.

5. a Table 710.1(2) 2000 dfu \div 5 floors = 400

6. c Table 710.1(1) 120 dfu \times 2 branch intervals = 240 dfu

7. c Table 710.1(1)

8. b Table 710.1(1)

Notes

Part 4: IPC Practice Exam #5 Answer Key

1. a 802.2

2. d 802.4

3. d 903.2

4. d 916.3

5. c Table 916.5.1

6. a 908.3

7. a 917.4

8. b 905.4

9. c 803.2

10. b 901.2

11. b 911.1 and Chapter 2 (Definition)

12. c Table 916.1

13. a 908.1

14. b 916.4.1

15. a Table 906.1

16. a 802.3.2

17. c Table 909.3

18. d 914.2

19. c 917.3

20. d Table 909.3

21. d 802.2.1

22. a 904.6

23. c 902.2

24. c Table 906.1

25. a 916.1

Notes

Part 4: IPC Practice Exam #6 Answer Key

1. d 1002.4
2. a Table 1102.5
3. a 1106.3
4. c 12,000 sq. ft. ÷ 6 leaders = 2,000
5. d 1113.1.2
6. a 1113.1.4 Exception
7. d 1002.3
8. d 1101.2
9. b 1109.1
10. d 1105.1
11. c 1002.1
12. a Table 1003.3.4.1
13. b Table 1106.2
14. c 1111.1
15. d 1113.1.4
16. a 1110.4
17. d 1111.1
18. b 1003.4.2.2
19. d 1002.2
20. a 1003.5
21. b 1103.3
22. c 1003.4.2.1
23. d Table 1106.3
24. b 1110.4
25. c 1106.3

Notes

Part 4: Storm Drain Diagram Answer Key

Leader Size: 5"

Horizontal A: 6"

Horizontal B: 8"

Horizontal C: 8"

Horizontal D: 10"

Notes

Part 4: Plumbing Isometrics Answer Key

	# Fixtures		dfu Value		Total
Water closets	8	×	6	=	48
Urinals	2	×	4	=	8
Lavatories	16	×	1	=	16
Floor drains	7	×	2	=	14
Shower heads	20	×	2	=	40
Water fountains	2	×	½	=	1
Washing machines	5	×	3	=	15
Laundry tray	1	×	2	=	2
				Total =	144

Part 4: Plumbing Isometrics
Answer Key (cont.)
Horizontal Drain Sizing (Figure 3)

Tip: When calculating a drainage fixture unit total for a horizontal drain, make a list of each type of fixture on the branch. This will help ensure all fixture types are accounted for. From this point, an accurate tally of drainage fixture units can be determined to correctly size the horizontal drain.

	Fixture Type	**# of**		**dfu Value**	**Total**
Horizontal Branch A:	Water Closet	3	×	6	18
	Floor Drains	2	×	2	4
	Urinals	2	×	4	8
				=	30

Horizontal branch A has a total of 30 dfus. Using Table 710.1(2) an accurate drain size can be determined.

What is the minimum drain size for horizontal branch A? **Ans:** *4"*

Use Figure 3 on page 91 to successfully complete the rest of the worksheet.

	Fixture Type	**# of**		**dfu Value**		**Total**
Horizontal Branch B:	Lavatories	8	×	1	=	8
	Floor Drains	1	×	2	=	2
	* Shower Drains	4	×	5	=	20
				Total	=	30

* *Shower Drains = # of Shower heads ÷ # of floor drains in shower area*

Shower drains and floor drains have a different dfu value because shower drains are required to discharge a larger quantity of water.

	Fixture Type	**# of**		**dfu Value**		**Total**
Horizontal Branch C:	Shower Drains	4	×	5	=	20
	Fountains	2	×	½	=	1
	Floor Drains	1	×	2	=	2
	Laundry Tray	1	×	2	=	2
				Total	=	30

Part 4: Plumbing Isometrics
Answer Key (cont.)

	Fixture Type	# of		dfu Value		Total
Horizontal Branch D:	Washing mach.	5	×	3	=	15
	Floor Drains	1	×	2	=	2
				Total	=	17
Horizontal Branch E:	Water Closets	5	×	6	=	30
	Floor Drains	1	×	2	=	2
				Total	=	32
Horizontal Branch F:	Lavatories	8	×	1	=	8
	Floor Drains	1	×	2	=	2
				Total	=	10

Each branch dfu total can now be used to size the horizontal drains.

1. What is the minimum drain size for horizontal branch B? 4"

2. What is the minimum drain size for horizontal branch C? 4"

3. What is the minimum drain size for horizontal branch D? 3"

4. What is the minimum drain size for horizontal branch E? 4"

5. What is the minimum drain size for horizontal branch F? 2½"

Sizing Building Drain (Figure 3)

When sizing a specific section of a building drain, one must first determine the total number of drainage fixture units flowing into that particular zone of pipe. Using Table 710.1(1), choose the correct column according to the given slope and size that specified section of drain pipe.

Building Drain A–B: This zone is conveying all fixture units discharged from horizontal branch A: (30 dfus). The slope, given in Figure 3, is ¼".

Column 4 of Table 710.1(1) determines the maximum number of dfus allowed to connect to a certain diameter of pipe having a ¼" slope.

Part 4: Plumbing Isometrics
Answer Key (cont.)

	Total dfu Value		Building Drain Size
Building Drain A–B:	30	=	3"
Building Drain B–C:	60	=	4"
Building Drain C–D:	85	=	4"
Building Drain D–E:	102	=	4"
Building Drain E–F:	134	=	4"

Vent Sizing (Figure 3)

IPC 2009, Table 916.1: This is the most common table used for sizing vents. The drain size and the number of fixture units being vented must be known in order to properly size the vent.

As if you notice, in the first column, there are multiple listings of the same diameter of pipe. To ensure the correct diameter is chosen, the total fixture units being vented must be selected from column 2. The next step is to determine the developed length of vent pipe, which is usually given in the figure illustrated or the question. Everything needed to determine the vent size is in Table 916.1. Make sure you pay close attention to what each column represents.

	Drain Size	# FU Vented	Developed Length		Vent Size
Horizontal Branch A:	3"	30	30'	=	2"
Horizontal Branch B:	4"	30	30'	=	2"
Horizontal Branch C:	4"	25	20'	=	2"
Horizontal Branch D:	3"	17	30'	=	1½"
Horizontal Branch E:	4"	32	30'	=	2"
Horizontal Branch F:	2½"	10	30'	=	1½"

Notes

Part 4: Isometric Practice Exam Answer Key

Diagram A:

1. d
2. b

Diagram B:

1. a
2. a
3. c

Diagram C:

1. d
2. a

Diagram D:

1. d
2. c

Diagram E:

1. b
2. a
3. c
4. b

Notes

Part 4: Isometric Practice Exam Answer Key (cont.)

Diagram F and F1:

1. b
2. d
3. a
4. d
5. d
6. b
7. c
8. a
9. d
10. b
11. c
12. a
13. d
14. c
15. b
16. d
17. b

Notes

Part 5: IPC Final Review Exam Answer Key

1. c Table 608.15.1
2. a 604.9
3. d 312.7
4. d 708.3.3 (must be greater than 45°)
5. b 502.3
6. d 405.3.1
7. c Table 308.5
8. b Table 608.17.1
9. c Table 709.1
10. a 802.4
11. d Table 403.1
12. b 417.3
13. d 604.5
14. b Table 710.1(2)
15. b 603.1
16. c 1107.3
17. a 604.8
18. d Table 916.1
19. c 312.3
20. b 306.3
21. d 403.2 exception #2
22. b 405.3.1
23. a Table 709.1
24. a Table 909.3
25. d 302.1

Part 6: IFGC Practice Exam Answer Key

1. a 404.10.1

2. b Table 403.9.2

3. c Table 402.2

4. a 403.4.2

5. d 403.13

6. c 403.10.2

7. c 411.1.3.1

8. b Table 415.1

9. c 410.2

10. b 302.3.2

11. a 413.8

12. a Table 402.2 and Table 402.4(2)

13. a 402.6

14. c 302.3.3

15. c Chapter 2 definitions

Notes

Part 6: Gas Pipe Sizing Questionnaire Answer Key

1. d
2. c
3. c
4. d
5. c
6. d
7. c
8. b
9. b
10. b

Notes

Part 7: General Knowledge Practice Exam #1
Answer Key

1. d
2. c
3. d
4. b
5. a
6. c
7. d
8. c
9. d
10. b
11. c
12. c
13. a
14. a
15. c
16. c
17. c
18. c
19. c
20. a
21. b
22. d
23. a
24. b
25. b

Notes

Part 7: General Knowledge Practice Exam #2 Answer Key

1. d
2. e
3. b
4. e
5. c
6. a
7. d
8. d
9. c
10. b
11. d
12. c
13. b
14. c
15. b
16. a
17. a
18. d
19. d
20. b
21. c
22. a
23. a
24. c
25. b

Notes

Part 7: General Knowledge Practice Exam #3 Answer Key

1. b
2. a
3. c
4. b
5. b
6. b
7. c
8. c
9. a
10. b
11. b
12. d
13. c
14. a
15. b
16. d
17. a
18. c
19. c
20. a
21. d
22. b
23. b
24. a
25. a

Notes

Part 7: General Knowledge Practice Exam #4 Answer Key

1. c
2. b
3. a
4. d
5. a
6. c
7. d
8. a
9. c
10. d
11. b
12. b
13. a
14. b
15. d
16. b
17. d
18. a
19. c
20. d
21. a
22. a
23. b
24. a
25. c

Notes

Part 7: General Knowledge Practice Exam #5 Answer Key

1. d
2. d
3. c
4. a
5. d
6. d
7. c
8. b
9. b
10. d
11. a
12. b
13. b
14. d
15. d
16. c
17. d
18. d
19. a
20. d
21. c
22. c
23. a
24. a
25. c

Notes

Part 7: General Knowledge Practice Exam #6 Answer Key

1. b
2. c
3. a
4. c
5. d
6. a
7. c
8. b
9. d
10. d
11. b
12. a
13. a
14. b
15. c
16. d
17. a
18. a
19. b
20. b
21. b
22. d
23. c
24. b
25. a

Notes

Part 7: "Identify" Quiz Answer Key

1. a
2. b
3. c
4. c
5. b
6. d
7. d
8. d
9. d
10. b
11. c
12. d
13. a
14. b
15. d
16. c
17. a
18. a
19. c
20. b
21. d
22. c
23. a
24. c
25. a

Notes

Part 8: Project Management Practice Exam #1 Answer Key

1. a
2. c
3. c
4. a
5. c
6. d
7. c
8. a
9. c
10. c
11. d
12. a
13. c
14. d
15. b
16. d
17. d
18. b
19. d
20. c
21. a
22. a
23. a
24. a
25. a

Notes

Part 8: Project Management Practice Exam #2 Answer Key

1. b
2. a
3. b
4. c
5. a
6. b
7. c
8. a
9. b
10. a
11. b
12. a
13. b
14. b
15. c
16. c
17. a
18. b
19. b
20. b
21. b
22. d
23. a
24. a
25. a

Part 9: OSHA Practice Exam Answer Key

1. c 1926 Subpart P, Appendix B (Table B-1)
2. b 1926.1053 (a) #13
3. b 1926.302(a)(2)
4. c 1926.651(J)(2)
5. b 1926.102
6. b 1926.24
7. c 1926.651
8. d 1926.651(h)(1)
9. a 1926.56 (d)(3)
10. a 1926.50 (d)(2)
11. c 1926. 651(c)(2)
12. c 1926.302(b)(4)
13. c 1926 Subpart P, Appendix A
14. a 1926.55 (a)
15. a 1926.501(b)(1)
16. d 1926.300(d)(3)
17. d 1926.25(b)
18. d 1926.1053(b)
19. d 1926.651(c)(2)
20. c 1926 Subpart P, Appendix A
21. b 1926 Subpart P, Appendix A
22. b 1926.1053(b)(1)
23. b 1926 Subpart P, Appendix B
24. b 1926 Subpart P, Appendix A
25. d 1926.28

Notes

Part 10: Backflow Prevention Answer Key

1. a
2. d
3. b
4. a
5. b
6. b
7. c
8. c
9. b
10. b
11. a
12. a
13. a
14. b
15. c
16. c
17. b
18. c
19. b
20. a
21. b
22. d
23. d
24. a
25. a
26. b
27. d
28. c
29. d
30. b
31. c
32. b

Notes

Part 10: Backflow Prevention Answer Key (cont.)

33. d

34. d

35. c

36. b

37. b

38. a

39. a

40. b

41. c

42. d

43. c

44. d

45. d

46. c

47. d

48. c

49. d

50. d

51. b

52. d

53. b

54. a

55. c

56. d

57. d

58. b

59. c

60. b

61. c

62. d

63. b

64. a

Notes

Part 10: Backflow Prevention Answer Key (cont.)

65. b

66. a

67. b

68. c

69. b

70. a

71. c

72. d

73. a

74. d

75. b

76. a

ABOUT THE AUTHOR

American Contractors Exam Services

American Contractors Exam Services has successfully prepared candidates for state-required licensing exams since 1991. A team of full-time instructors comprises field experience, training expertise, and formal education to formulate an effective and dynamic approach to concisely present plumbing codes, math concepts, and plumbing theory in a seminar format. The seminars are conducted throughout the United States utilizing multimedia and interactive technology.

Utilizing effective graphics, pictures, and bulleted text, complex material is presented in an easy-to-understand format. This simplified approach has led to thousands of success stories and happy clients. One is sure to benefit from any material written or presented by American Contractors Exam Services.

Many of the illustrations featured in this book are used in seminar presentations throughout the country by American Contractors Education Services.